A Vogt
for the
Environment

The true story of how one teenager, Tanja Vogt,
convinced McDonald's to show concern
for the environment

enviro-kids

by John Sailer

jsailer@aol.com

The Book Publishing Company
Summertown, Tennessee

Cover and interior design by Barbara McNew
Front cover photo by Lorin Klaris
Back cover photo by K. Vogt

Special thanks to our junior editorial staff:
John Schweri, Sophia Casini, and Emily Cook

Library of Congress Cataloging-in-Publication Data
Sailer, John, 1963-
    A Vogt for the environment / John Sailer.
    p. cm.
    Summary: Relates how teenaged environmentalist Tanja Vogt
convinced McDonald's, as well as businesses and schools in her
area, to stop using styrofoam.
    ISBN 0-913990-34-5
    1. Vogt, Tanja–Juvenile literature. 2. Environmentalists–United
States–Biography–Juvenile literature. 3. Plastic scrap–Juvenile
literature. 4. Food containers–Juvenile literature. 5. McDonald's
Corporation–Juvenile literature. [1. Vogt, Tanja, 2. Environmentalists. 3. Environmental protection. 4. McDonald's Corporation.]
I. Title.
    TD170.V64S25 1993
    363.7'0092–dc20
    [B]   93-15313
                                                  CIP
                                                  AC

ISBN 0-913990-43-5

0 9 8 7 6 5 4 3 2 1

Contents

Chapter 1   *See me after class* 5

Chapter 2   *Put your money where your mouth is* 15

Chapter 3   *A nickel a day for paper trays* 22

Chapter 4   *The anit-styrofoam campaign spreads* 29

Chapter 5   *McDonald's not easily swayed* 36

Chapter 6   *Tanja addresses the United Nations* 45

Chapter 7   *An education in plastics* 48

Chapter 8   *Disorder in the court* 56

Chapter 9   *Publicity brings fame* 62

Chapter 10   *Earth Day* 69

Chapter 11   *Sticking her neck out and into Russia* 79

Chapter 12   *McDonald's makes an announcement* 83

Appendix:

           Organizations 87

           Miscellaneous 91

           Books & Magazines 91

           Awards 95

4

Acknowledgements

Thanks for the completion of this book go to Tanja Vogt herself, of course, and all of the other enviro-kids whose clear vision and dedication enabled them to achieve their goals; Tanja's parents, who supported her (and continue to do so) throughout her entire campaign; Karl Stehle and Norma Stehle (formerly Sullivan), the teachers, now husband and wife as well, who guided these young environmentalists; Karen Westdyk, who put me in touch with a publishing company that, like herself, exists for people and the environment as well as for the purpose of publishing; and my wife, Eileen, for her encouragement.

Chapter 1

See me after class

"See me after class," said the note on Tanja's paper. It was Monday, first period, and the teacher, Mr. Stehle, had just handed back his history class's current events assignments from the previous Friday.

"Oh no! What could he want?" Tanja wondered immediately after reading the note written just below the grade. Both the grade and the note were written in red ink, the preferred medium of all teachers, elevating the urgency and importance of his comments. "See me after class," Mr. Stehle's note ordered from the white page.

"Whatever he wants can't be that bad," Tanja reassured herself. "He gave me a ✓+," meaning that her homework was above average. Still, the teacher's note formed a dark cloud of fears that rumbled through Tanja's mind: "He disagrees with my opinion. He thinks I should work harder. He wants me to do extra credit. He wants something from me, but what? What could he want?" The cloud of fears only got larger and blacker as the class period wore on.

Tanja couldn't concentrate on the class. All she could think about were Mr. Stehle's note and its mysterious meaning. She was lost in conjecture, anxious for the class to end. She glanced back and forth from the note to the clock on the wall, its red second hand sluggishly ticking out the minutes that would bring the class to an end.

"Tanja," the teacher said directly to her. His voice broke her trance. "Can you tell the class why the Sons of Liberty staged the Boston Tea Party?" His question was based on the discussion he was conducting right at that moment in the classroom, and Tanja had not been paying attention.

"What?!" she thought, surprised. For a split second she was tongue-tied by his unexpected question. She felt her classmates' eyes upon her. Time slowed further, nearly stopping, and she was eternally trapped in the converging stares of her teacher's and her classmates' eyes. But even though she had been distracted in class and had no idea what discussions had led to this question, she was prepared because she had read the appropriate chapter in her history book. Although startled, she could answer the question.

"They were protesting the taxes imposed by the British," Tanja answered, simple and straightforward, no need for more. She didn't offer much detail because she didn't want to repeat anything that might have been said earlier while she had been lost in meditation, staring at the note emblazoned in red ink on her homework paper.

"Exactly right," said Mr. Stehle with enthusiasm and a smile that praised Tanja's response. "Now, what were the results of these protests?" he continued. But his new question was meant for someone else in the class, and Tanja was again distracted by the red note on the paper in front of her. "See me after class," it commanded with self-imposed importance.

"I *will* see you in ten minutes," she thought to herself, silently answering the note and adding up the minutes until the end of class.

"It can't be anything bad," she thought, trying to reassure herself again but having little success. "He just smiled at me and said, 'Exactly right' when I answered his question. He called on me when I didn't even raise my hand. He must have thought I knew the answer, unless he thought I was daydreaming and wanted to catch me. I'm sure the note can't mean anything bad. Maybe he just wants to tell me something about the newspaper article I handed in. Maybe I didn't write enough in my essay about the article. Maybe he didn't like what I wrote."

Bbbrrring! Finally the period was over. Most of Mr. Stehle's first-period history class had already lined up at the door, books in hand. At the sound of the bell, they sprung into the instantly crowded hallway and stampeded to their next class.

Tanja lagged behind, dawdling over her books until the classroom had emptied of all but herself and her teacher. Then she approached Mr. Stehle.

"You wanted to see me?" Tanja asked, tentatively holding

her paper out over his desk, pointing to the note he had written on her homework assignment.

"Tanja," said Mr. Stehle, looking up from the worn leather briefcase he was stuffing with papers. He looked her straight in the eye and asked, "Why don't *you* do something about it?" Then he looked back down at his briefcase acting as if she knew exactly what he was talking about.

His abruptness took her by surprise, and although he was referring to the homework paper she held in her hand, she didn't understand what he was talking about. "What? What do you mean?" she asked, furrowing her forehead and raising one corner of her mouth in a questioning half smile. "I did do something about it. I came to see you as soon as I could."

"Not about my note," he said, laughing. "About the article you handed in for current events. You said something should be done about this. Why don't *you* do something about it?" he asked for the second time. Now she did understand what he was talking about.

"Oh, the newspaper article. You want *me* to do something about it? What could I do?"

"Anything you want," he said.

"I couldn't do anything about it."

"Of course you could," said Mr. Stehle, "if it matters enough to you."

And it did matter to Tanja. That's exactly why she had selected the article in the first place.

It had been the previous Thursday, the night before her weekly current events assignment was due, when Tanja scanned the paper to choose an article.

"Nothing ever happens in this town," Tanja thought as she flipped through the local newspaper spread out on the carpeted floor of her bedroom. The large pages crackled as she turned them. The 15-year-old didn't normally make a habit of reading the paper, except on Thursday nights when her current events assignment was due the next day and sometimes on Sundays when she read the comics and the advertising inserts. Even now for current events she only scanned the headlines of the October 17, 1988, issue of the *West Milford Argus*—"Residents oppose quarry permits" and "VFW post to be rededicated" and "Neighbors concerned about foster home" and other stories that could pertain to any small town throughout the country. It was a challenge to find something unique, something interesting, something important.

"Boring," she thought after reading the headlines on the first few pages. "Who cares whose birthday it is?" she wondered as she turned past an entire page devoted to pictures of kids, and even some adults, celebrating birthdays that week. Having lived in the same house in the same small town all of her life, she recognized many of the people pictured on the page. "This paper's hard up for news," she thought as she turned past the page of birthday announcements. "What will I hand in for current events this week?"

"Choose a subject that you care about," Mr. Stehle had instructed Tanja and her classmates when earlier that day he had reminded them of their weekly assignment. Every Friday, the students in Mr. Stehle's history class had to hand in a newspaper article accompanied by a short essay summarizing the article and describing their opinions. Every Thursday, when he reminded them of their weekly assignment, he said the same thing—"Choose a subject that you care about."

His voice echoed silently in Tanja's brain as she scanned the headlines for a second time, but all the articles seemed to be about local politics and small town budget discussions, neither of which interested Tanja in the least. Even West Milford's crime was boring—the brief articles that described the few drunken driving arrests, the listings of criminal mischief that reflected the previous evening's innings of rural delivery mailbox baseball, and the overblown domestic disputes were unfortunately not unusual enough to warrant using them for current events.

"Doesn't anything ever happen in this town?" Tanja wondered as she sat with her back against the soft comforter of her neatly made bed and continued scanning the local biweekly newspaper spread out before her on the floor. Then, a headline caught her eye. It was buried in the middle of the newspaper.

"Despite threat to all life, board continues to use styrofoam trays," said the headline.

"Threat to all life?!" thought Tanja, wondering what decisions could possibly be made in her small New Jersey town of West Milford that could threaten all life. Intrigued by the headline, she read the entire article:

"**WEST MILFORD** — Township Board of Education members expressed concern over the continued use of styrofoam trays for school lunches.

Recently, the board decided not to enter into a contract for either paper or styrofoam trays. They decided to purchase the trays as they are needed so they can buy paper trays as often as the price allows.

The problem with paper trays is that they are much more expensive than the ones made of styrofoam, school officials said.

The board's Business Administrator, Edward Vogel, explained that if the school were to purchase paper trays for the entire year at the current price, they would have to raise the school lunch price by five cents. He said that each time the price is raised the number of students who purchase lunch decreases by between two and three percent.

Board member Trisha Banis said that she is concerned about the continued use of styrofoam because of the hazard to the environment. She explained that a gas, chlorofluorocarbon or CFC, that is given off by styrofoam when it breaks down is dangerous to the ozone layer.

CFCs, which are also used as a refrigerant, break down ozone into simple oxygen molecules. Ozone in the stratosphere blocks hazardous ultraviolet rays from entering the atmosphere. Without the ozone layer, scientists believe all life on the planet would perish.

She thinks the board should consider the five cent raise so that they can purchase paper trays. Several other board members agreed that they were concerned with the environmental hazards of the styrofoam.

Banis also asked about the possibility of using washable plates and trays. Vogel explained that dishwashers were discontinued during the energy crisis. Most of the schools no longer have dishwashing equipment. He added that it would be very costly to hire a dishwashing staff.

Board president Carl Richko suggested purchasing the styrofoam for now and reviewing the situation in December. He said that he would rather use paper but agreed with Vogel that it is currently too expensive."

" 'Currently too expensive'?" thought Tanja, amazed at their reasoning. " 'Currently too expensive'? Yeah, right. We'd better save our money now and kill ourselves in the process just to be sure we have more money when we're dead," Tanja thought sarcastically to herself. "Everybody knows that the person who dies with the most money wins."

The thought reminded her of a joke: "Your money or your life," said the man with the gun. "You better take my life," answered the victim. "I'll need my money for my old age." To Tanja, the board's decision to buy styrofoam trays seemed just as foolish.

"Now that's something I care about," Tanja thought after she read the article. "Something should be done about this. What's more important, money or the environment? We could all die without the ozone layer. Isn't that worth five cents?"

Unaware of the challenges and fame this simple homework assignment would bring her, Tanja cut out the article and turned to a fresh, white piece of lined paper in her notebook. Sitting once again with her back up against the soft bed, she began drafting the essay that would complete her current events assignment.

She began with five false starts and two rewrites, and although she had written little on each of the unfinished pages, she tore them from her notebook, crumpled them into balls, and played basketball with the wicker wastebasket in the corner. Finally, Tanja completed the brief essay that would accompany her current events article, clipped it to the article itself and tucked it into a glossy blue folder marked "History." The balls of crumpled paper gathered in and around the wastebasket were evidence that although the teenager was concerned about the "threat to all life" reported in the paper, she was still unaware that even her own smallest actions had an impact on the environment. She ignorantly wasted paper because she hadn't yet learned of the importance of a slogan that would soon become her rallying cry—"Reduce, recycle and reuse."

Tanja stacked her history folder atop the other school books and papers on her desk, gathered up the crumpled balls of paper that had missed the wastebasket and, still unaware of her wastefulness, threw them into the garbage. Now free from the responsibility of doing last-minute homework, Tanja laid down on her bed and continued reading *The Good Earth* for

English class. She didn't think of the current events assignment until the next day when Mr. Stehle asked the class to pass their homework forward. Tanja obliged, and that was that, ... or at least that was what she thought until the day Mr. Stehle handed back her paper with the note, "See me after class."

◆ ◆ ◆

"So, what do you think I could do?" Tanja asked her teacher after she realized that removing styrofoam from her school really did matter to her.

"You could tell the school board how you feel," said Mr. Stehle. "Maybe you could even get them to reverse their decision."

"Yeah, right. They wouldn't listen to me. I'm just a high-school kid. All they care about is money, and they're not willing to spend more for paper trays, even if they are better for the environment. They say they're too expensive."

"Well, you said in your essay that you'd be willing to pay more. Do you think the other kids would also pay more for lunch if they knew paper trays were better for the environment?" Mr. Stehle asked.

She thought for a minute, first trying to understand her own feelings before she presumed to understand how any of the other kids might feel. "Yeah, I guess I would pay more. Maybe the other kids would, too. I don't know. I'm not sure."

That was just the answer her teacher had hoped for. "Why don't you find out? Conduct a survey. You could ask the other kids if they would be willing to pay more, and if they would, you could tell the school board. Don't you think the members of the board would listen to the kids? You're the people they're working for in the first place."

"Maybe. I don't think it would matter, though, because all they care about is money. It's right here in the article. They already discussed the environmental issue but decided that keeping the price to a minimum was more important than keeping the environment clean. Even the headline says 'Despite threat to all life, board continues to use styrofoam.' I don't see the point in telling them that I would be willing to pay more for paper trays when they already considered that alternative," Tanja concluded. In part she was justifying her fear of presenting her feelings before the school board, but she was also thinking, "This was a homework assignment not a crusade against styrofoam, right?"

"They might listen if it wasn't just you who told the board that you'd be willing to pay extra for the paper trays," Mr. Stehle said. "What if the other students agreed with you and told the school board their feelings, too? Give the school board a list of hundreds of signatures of students who feel the environment is more important than saving a little money.

"What do you think the Sons of Liberty would have done?" continued Mr. Stehle. "Don't think that when I asked you about the Boston Tea Party in class today I wasn't thinking about your current events assignment and how you disagreed with the board. Do you think the Sons of Liberty should have said, 'Oh well, King George has considered the possibility of not taxing us, but since he decided that he'd rather have more money, he won't even realize how we feel if we trash the shipments of the tea he's trying to tax.'? They wouldn't have been standing up for their beliefs if they had decided that and gone on obediently paying their taxes. We'd still be a British colony if nobody disagreed with authority. And today we might be the last 'colony' on earth if nobody stands up for the environment now and disagrees with the use of styrofoam."

"I don't think you can compare this with the Boston Tea Party," said Tanja, who still was not convinced, although she actually found the idea of leading a crusade against styrofoam more inviting after Mr. Stehle put it into this historic context. "Hmmm, the West Milford Styrofoam Lunch Tray Party," thought Tanja, half kidding herself but also half serious. "It could go down in history," she thought, allowing herself to become impressed with her insight for finding probably the one thing in the entire *West Milford Argus* that had such a universal scope.

"You *can* compare it to the Boston Tea Party," he said, "if you see both as simply standing up for something you believe in, for questioning authority when authority is wrong. You believe that paper trays are the more environmentally considerate decision, even though they cost more, and the facts support your belief. I believe that the school board will listen to you if you tell them how you feel. You say that something should be done about this, and I agree, so why don't *you* do something about it?" Mr. Stehle asked Tanja for the third time since she had approached him with her current events assignment.

"Do you think I should? Do you agree with me? Do you think the board will change?" Tanja asked, looking for some support to help her make her decision, leaning toward accepting the challenge now that their discussion had transformed it into a historic cause.

"I don't want you to do anything because of what I believe. This should be your decision. But let's just say that there are a lot of important things that can't be measured in dollars, and the environment is one of them."

"Well, that sounds like he agrees with me," thought Tanja.

She thought about Mr. Stehle's belief in her, the value he had for standing up for one's beliefs, and her sincere concern for the environment that first led her to choose the "Despite threat to all life" article for current events.

"Maybe I could stage my own Boston Tea Party in West Milford," she thought to herself, "Yeah, we could dump all the used styrofoam lunch trays on the school board president's front lawn until the board changes their minds. And even if the board doesn't change, at least I'd know that I had expressed my opinion." And in her mind, the line, "Something should be done about this," that less than one week before had simply been a good way to end a current events essay, led her to answer her history teacher, "I guess I could try to do something about it. What do you think I should do first?"

Mr. Stehle smiled, then firmly grabbed her shoulders and gripped them in such a way that said, "I'm proud of you, that's what I hoped you would say."

Praising her as if she had just scored a point in some sporting event, he said, "Alright, Tanja! That's great! How about we first find out how your classmates feel, and if they agree with you, then maybe we can conduct a survey of the whole school. If enough students agree with you, I'm sure the school board would reconsider and at least resume talking about the possibility of switching to paper trays."

"But I've never done anything like this before. I'm terrible at speaking in front of a group, and I don't know how to conduct a survey. Would you help me out?"

Mr. Stehle smiled again, which, along with his curly brown hair and bushy moustache, softened his face and simultaneously encouraged and comforted Tanja as she embarked on her campaign. But her teacher was even more excited. He would have gotten started that minute if he didn't have a day's worth of classes to teach. He had always emphasized applying his lessons to real life, and now he had the chance.

*Tanja and current events' teacher Karl Stehle at a
McDonald's boycott rally.*

Chapter 2

Put your money where your mouth is

Later that day, at lunch, Tanja stood in the cafeteria line watching the students ahead of her as one after the other took a styrofoam tray from the tall stack to the left of the bright stainless steel service counter. None of the other kids even thought twice about the impact their actions had on the environment. Few, if any, even knew of the material's "threat to all life," and Tanja herself only knew of its dangers because she had been required to prepare a current events assignment. Less than one week before, she had been as ignorant as they were.

Tanja hesitated after the line had moved forward and it became her turn to take a tray. She didn't want to. Instead, she wanted to grab the trays from everyone ahead of her and tell them all how harmful the styrofoam was for the environment. Then she wondered how all of the other students would react. She thought that they might all laugh or just ignore her. So she took a tray anyway. Simply holding the dry, porous plastic made her feel guilty.

After Tanja sat down among her girlfriends and lunchtime acquaintances in the cafeteria, she wondered if they would really agree to pay more for lunch every day just to save the environment. She started to have second thoughts as she listened to her friends' conversations about clothes, boys, school, movies, and TV. "Maybe they won't pay extra to help the environment," she thought. "Maybe it doesn't really matter to them. Should I ask them?" she wondered to herself. "Should I just ask them right now to see what they really think before I go through with this? Maybe they don't even care."

Lunch was about half over before there was a lull in the girls' conversations and before Tanja had gathered enough courage to ask her companions what they thought.

"Do you know that these lunch trays are bad for the environment?" Tanja finally ventured, tossing her question into the brief silence of the group.

"What?!" said Jackie, surprised by Tanja's out-of-the-blue remark.

"What are you talking about?" asked Kerry.

"No, really," persisted Tanja. She brought it up, and now she had to finish it. "These trays are made of styrofoam, and styrofoam is bad for the environment. It harms the ozone layer, and the only reason the school board decided to use these instead of paper trays is because of money. Styrofoam's cheaper."

"What's the endzone layer?" asked Jackie.

"The ozone layer," corrected Tanja, "Don't you even know ... ?"

"Oh, I know what it is," Jenn volunteered even before Tanja had a chance to continue. Relieved by the support, she briefly relaxed in her orange, molded plastic chair. "It's a layer in the atmosphere that protects us from the sun," Jenn continued, "but there's a big hole in it over Antarctica, and it's getting bigger all the time. The more it disappears, the more people are going to get sunburns and skin cancer and even cataracts because the ozone won't be there to protect us from the sun's ultraviolet rays. But I thought the hole was from aerosol cans and air conditioning and refrigerators, not from styrofoam."

"I don't know anything about all that other stuff," said Tanja, who still only knew as much as she had learned from the single *Argus* article she had brought in for current events, "but I know that styrofoam is bad for the ozone layer. Would you guys be willing to pay more for lunch if it would help protect the environment?"

"I would," said Jenn immediately. "I already stopped using aerosol deodorant and hairspray."

"It depends on how much more it would cost," said Jackie.

"Yeah, I might too, depending on how much it would cost," agreed Kerry.

"Always a price tag, even on the environment," Tanja thought to herself. But she didn't dare say what she was thinking for fear of being labeled a "dweeb." "Just like the board of education."

The next day, the reaction from the students in Mr. Stehle's history class was not only less enthusiastic than that of Tanja's lunch companions, it was nearly nonexistent. After Mr. Stehle described Tanja's article to the students in his first-period history class, he said, "This can't be Tanja's thing alone. It would have to come from the class. It's too big. She said that she thinks you'd be willing to pay an extra nickel. What about it?"

No response. Blank stares.

Mr. Stehle kept talking and tied the idea into his continuing lesson on the American Revolution, except this time it was Thomas Paine and his *Common Sense* pamphlet that he compared to Tanja's idea. Paine had written *Common Sense* and distributed it among the colonists to encourage support for a peaceful revolution against England. Mr. Stehle suggested that the kids could write their own pamphlet about the perils of styrofoam and the reasons for switching to paper. He also suggested that they could start a petition or go to the school board. In fact, he spent the entire period talking about the possibilities for convincing the school board to reverse its decision. Since the students offered little response, Mr. Stehle did most of the talking. He even warned them that they might not even want to get involved in a "styrofoam revolution" for fear of looking like fools among their friends.

Finally, he left it up to them. "What would you like to do about it?" he asked.

They were still nervous and made very little eye contact with Mr. Stehle. Even Tanja, who felt as if she had been placed squarely on the spot, did not respond. The bell was about to sound, and Mr. Stehle felt like he was playing poker. "I don't know what you're holding," he announced to the class. "You may have a full house or you may have a pair of deuces." They smiled a little, ... finally. "Let me know tomorrow. It will be your thing. I'll help you, but I'm not doing it. Let me know tomorrow." He had no choice but to wait for them to respond.

●◇   ●◇   ●◇

"So? What are you going to do?" Mr. Stehle asked his students at the start of class the next day. Based on their lack of response the day before, he anticipated little.

"We're ready to do it," several students agreed, surprising Mr. Stehle. Their blank stares of the previous day had unexpectedly

transformed into excitement. After discussing it among themselves the day before, the students had decided to accept the
challenge. They spent the entire class eagerly considering ways
to get the school to switch from styrofoam to paper.

Mr. Stehle offered his own suggestions. "Paine would have
organized a lunch boycott," he said, maintaining his references
to the American Revolution to ensure that the discussion never
really strayed far from the legitimate curriculum.

The students had their own concerns. One said what they
were all thinking, "What if we're the only ones to show up with
lunch bags during the boycott? Everyone will laugh at us."

"Only one-third of the people favored the American Revolution," said Mr. Stehle in response, "but Paine's arguments for
independence converted many. You see the similarities? What
we face here with the environment is what they faced in the
revolution. It was a small core of people who believed they could
make a change."

After Mr. Stehle helped his students overcome their concerns that the other kids might vote against them or laugh at
them, they quickly got to work on their own "Common Sense"
pamphlet about abandoning styrofoam trays and switching to
paper. It became an ongoing classroom project.

With still more help from their teacher, they also made up
a survey asking the rest of the students in the school if they
would rather use a styrofoam tray for $1.20 or a paper tray for
$1.25. Of course, the survey included information about
styrofoam's harmful effects on the environment. The students
felt that this information would help the other kids make an
educated decision based on more than just the cost of the trays.
If it were just a question of money, the results would have been
obvious.

To gather the details about styrofoam, even Mr. Stehle had
to get help. And, just like his students, Mr. Stehle turned to a
teacher for the answers he needed. Norma Sullivan, a teacher in
West Milford's Macopin Middle School, was a close friend of Mr.
Stehle's and had a college degree specifically in environmental
education. She knew all about styrofoam's harmful effects and
was happy to share this information with Mr. Stehle when she
heard about his students' plans. In fact, the science and
language arts teacher's seventh and eighth graders were right in
the middle of studying the environment and the dangers caused

by chlorofluorocarbons, a by-product of styrofoam. It was this very same information Mrs. Sullivan was already using in her class that the students incorporated into their survey.

The one-page survey said:

Paper Trays vs. Styrofoam Trays

Issue: Styrofoam trays and cups are now being used in our cafeterias. We know that styrofoam is dangerous for the environment because:

   a. It releases harmful chemicals (chlorofluorocarbons CFCs) when burned which is causing the deterioration of our protective ozone layer surrounding the planet.

   b. It takes up unnecessary space in our overcrowded landfills.

   c. It is nonrecyclable and non-biodegradable.

Paper trays and cups cost more. Purchasing them will result in increasing the school lunches by 5 cents. Statistics have shown that cost increases in the past have contributed to a decline in buying lunches by two to three percent, according to business manager Mr. Ed Vogel.

At the bottom, the questionnaire asked, "Would you rather pay $1.25 for paper trays or $1.20 for styrofoam trays?"

Mr. Stehle was pleased to distribute the surveys among the students in his other classes, and the other teachers were equally willing to do the same.

The survey became a hot topic, not just in Tanja's first-period history class, but throughout the high school. It was discussed in the hallways, lunchroom, study halls, and gym classes. Even though most of the comments Tanja heard supported her belief that her classmates would pay more to protect the environment, she still doubted that the results would be in her favor.

Even Mr. Stehle had his own doubts, especially when the principal's only comment was, "This is cute, Karl."

However, it was neither cute nor doubtful when the results started coming back. Each class tallied up its own responses and sent the totals back to Mr. Stehle. Tanja's first-period history class, which had originally generated the idea, counted up the final results—of 603 students in the high school and middle school who answered the survey, 519 (86%) were in favor of using paper.

The kids were excited and sure that the school board had to switch now. They had no choice. The board couldn't go against the wishes of nearly all of the students in the high school. Could they?

"What now?" Tanja and her classmates asked of Mr. Stehle. By this time, a core group of environmentalists had formed in first-period history class. It was this group—Kurtiz Schneid, Monica Piergrossi, Athena Cosmas and Cavan Clinton—who, along with Tanja, saw the survey through to the end and now sought to pursue the campaign to its completion.

"Now you have to inform the school board," advised Mr. Stehle.

So they drafted a letter describing the survey and its indisputable results and sent it to the school board's business administrator, Edward Vogel. Then they waited for his answer. At this point, they had no doubt in the outcome. The school board would have to switch.

But it wasn't that simple. They waited a few weeks for the answer.

Then, on a Friday afternoon after the high school had already let out for the weekend, Ed Vogel called Mr. Stehle. "I got your information from the class," he said. "It's admirable, but they'll have to put their money where their mouth is."

"Excuse me, Ed?" asked Mr. Stehle.

"I've instructed the cafeteria personnel to give the students the option of getting their lunch on styrofoam for $1.20 or on paper for $1.25," said Mr. Vogel.

"What do you mean?"

"The students can prove that they truly are willing to pay more for paper trays by actually doing so. If they want plastic trays, they can pay $1.20 for lunch, but if they want paper, they'll have to pay $1.25," explained the school's business administrator. "This is their chance to prove that the environment is really more important to them than money."

"Okay, when do you plan to do this?"

"Monday morning."

"Monday morning, Ed?" asked Mr. Stehle, completely surprised. First of all, he never expected his students' survey to be challenged like this, and to do this without any notice was simply unfair. The high school had already let out for the weekend, and Mr. Stehle wouldn't see his students until Monday morning, too late to explain the procedure to the entire high school.

"You have to give me a chance to talk to the students and prepare," pleaded the teacher. But there was nothing he could do to sway Mr. Vogel, who was determined to test the students to prove whether they would really pay extra for paper trays.

Without preparation, Mr. Stehle was sure the students would choose styrofoam trays simply because they were cheaper. Their brief education in the dangers of styrofoam had been the surveys distributed by Tanja and her classmates, and he had seen enough test papers to know the reliability of a high school student's short-term memory. Few of the students would remember why they shouldn't use styrofoam and would simply vote with their wallets. He spent the weekend worrying that the students might make the wrong decision because there was no time to teach them the facts.

At the middle school, Mrs. Sullivan also heard about Mr. Vogel's plan. Her vice principal passed on the news. "Guess who just called," he said to her and then explained that the same test would be conducted in the middle school. "Next week will be the real test when they have to come up with the money. Good try." He assumed it was all over. He didn't believe the students would spend more on lunch just for something as abstract as a hole in the ozone layer.

But Mrs. Sullivan had more time to act than Mr. Stehle had. Even though she had been notified a little bit later that Friday afternoon, the middle school let out at 3:40 P.M., much later than the high school, which let out at 2:10 P.M.. She and her students had time to react. They filled the hallways and cafeteria with posters that promoted the use of paper and listed the pros and cons of using paper or styrofoam trays.

For Mr. Stehle, spreading the word was a little bit trickier. It had to wait until Monday morning.

Chapter 3

A nickel a day for paper trays

"I knew it," said Tanja when Mr. Stehle told her the bad news on Monday morning. "I knew they wouldn't listen to us, just a bunch of kids." From the time she had first read the " threat to all life" article in the newspaper until now, when Mr. Stehle told her the news that their questionnaires were not enough to convince the school board to reverse its decision, her relationship with styrofoam had changed from ignorance to disdain. She hated it, and now she wanted it out of her school.

"This doesn't mean it's over, Tanja," Mr. Stehle reassured her, maintaining his eternal optimism. "Mr. Vogel only said that the questionnaires weren't enough. That doesn't mean it's over. He wants us to prove it. He wants the kids to prove that they really would be willing to pay more for paper trays."

"Yeah, he probably just wants us to give up. He probably thinks that by making it harder for us, we'll just forget about it."

"I didn't say that." No, he didn't say that, but he thought it. "He wants us to prove it. Think of it as a challenge. Every problem has a solution, and we've got to find the solution to this problem. He wants the kids to literally put their money where their mouth is."

⊷　⊷　⊷

Luckily that day, the day Mr. Vogel's "put-your-money-where-your-mouth-is" test was set to begin, Mr. Stehle had an unexpected opportunity to teach the whole school about styrofoam's harmful effects. Just by chance, there happened to be an assembly for the entire school on drugs and alcohol that morning before lunch, and Mr. Stehle negotiated a few minutes to speak to the assembled student body.

Although he regularly spoke to classroom-sized groups of around 30 students, Mr. Stehle was not an experienced public speaker. He was nervous as he prepared to address the school's approximately 1,200 students, divided into two assemblies because of the limits of the auditorium. Although his palms were wet and his throat was dry as he walked up to the microphone on the stage, his passion overcame his fear and he rallied the students to the side of the environment.

" ... and Mr. Vogel said you can put your money where your mouth is!" Mr. Stehle called to the students after announcing the results of the survey and describing Mr. Vogel's plans to test them at the cafeteria cash register. The crowd went crazy, and Mr. Stehle waited for a lull in the screams and whistles before continuing to drive them into a pro-environment frenzy. He called again, "He wants to know, 'Are you willing to pay that extra nickel?'" And their screams in response assured Mr. Stehle that their votes would be for paper.

◆◇   ◆◇   ◆◇

During Trial Week, as Mr. Vogel's "put-your-money-where-your-mouth-is" test came to be called, Tanja and her classmates flooded the cafeteria with their anti-styrofoam "Common Sense" pamphlets and papered the walls with anti-styrofoam posters. And when some kids came unprepared to pay the extra nickel for paper, others were there to give them the money that enabled them to cast their environmental vote.

Tanja no longer had to feel guilty holding her tray in the cafeteria line. She chose paper. And now there was hardly a student in the entire school who was not as informed as she was about the effects of styrofoam on the environment. Even her lunchtime friends knew the difference between the endzone and the ozone and proved that they were willing to spring for an extra nickel every day at lunch to help protect the environment.

Still, Tanja wondered what the outcome would be. She spent a lot of time in the cafeteria that week, monitoring the test, and although she saw kids giving their friends nickels to pay for paper trays and overheard conversations in favor of the environmental stance she and her friends had taken, she heard many more conversations on completely unrelated topics.

It was the second week of December, and Trial Week actually seemed to be the last thing on everybody's minds. Even though there were anti-styrofoam messages posted all over the school, the teachers and the students were all preoccupied with other things—winter break, midterms, Christmas shopping, trips to warmer climates, everything but styrofoam and its effect on the ozone layer.

The only things Tanja's friends were talking about were the gifts they were buying or the gifts they expected to receive. And it got worse. Some students' attitudes toward paper trays had not only turned from sympathetic to apathetic but had completely reversed to adversarial. As Trial Week neared its end, there were some students who objected to the environmental peer pressure coming from the school's group of enviro-kids. They complained to Tanja and her classmates, who, in turn, brought the message back to Mr. Stehle, telling him that they were afraid their new opponents might even select styrofoam just for spite. "Back off," he advised. "You're learning what a revolution is. You've educated them. Now let them make their own decision."

And again, their decision was a successful surprise—72 percent of the high school students and 86 percent of the middle school students paid an extra nickel for paper trays. This time, Tanja and her classmates were sure the school board would listen.

⊶   ⊶   ⊶

With the results in their favor, the students were ready to confront the school board again, but this time it wasn't simply a matter of sending a letter and waiting a few weeks for a response. This time they planned to meet the board face-to-face at one of their monthly meetings and present them with the facts.

"There's nothing to be nervous about," Tanja's father reassured her as they drove together toward the school. The unpolluted sky above the mountain town of West Milford was full of stars, and the car was just beginning to warm up on this cold night only a few days before Christmas. But the holiday was far from Tanja's mind as she worried about the presentation she was about to make before the board of education.

"Calm down, Tanja," her father said for the umpteenth time. He was concerned that her shivering was now due more to nerves than the temperature.

"Calm down?" thought Tanja. "Easier said than done." She had never been to a board of education meeting and didn't know what to expect. She couldn't help picturing a panel of bulbous-faced men in powdered wigs and black robes sitting behind a towering wooden desk bellowing commands and hammering gavels. "They won't listen to me," she thought. "I'm just a kid."

Actually, the school board differed from her conjured image. The members, both men and women, were dressed in normal street clothes and sat at long tables arranged in a U-shape in the high school library. They sat unpretentiously in the same metal folding chairs that were also arranged in rows facing them for the small audience of spectators that generally gathered for school board meetings. Microphones were set on each table to record the official proceedings.

Still, even though there were no gavels, powdered wigs, or black robes, the room's layout and its implied formality did not ease Tanja's apprehension. She could hardly sit still in the front row seat she had taken alongside Mr. Stehle, who had already arrived and was waiting for the entourage of enviro-kids he had selected to speak that evening.

"I'll sit back here," Mr. Vogt whispered to his daughter as he crept toward a seat in the rear.

One by one, Tanja's classmates joined her in the front. There was Kurtiz Schneid, who was in Mr. Stehle's class with Tanja; Damian Cavalli, a senior who had helped with the anti-styrofoam campaign; David Corsaro, one of Ms. Sullivan's middle school students; and some other kids who had helped encourage the selection of paper trays in the cafeteria during Trial Week.

Although she and her classmates were first on the agenda, Tanja had to endure the role call, the approval of minutes from previous meetings, and other routines that seemed to take forever. She wondered when she would be called to speak and kept glancing at the clock like she had on the day that Mr. Stehle had handed her the note that commanded, "See me after class."

Finally, it was Tanja's turn. Mr. Stehle leaned forward and whispered words of encouragement as she picked up her notes with her sweaty hands. She could feel her heart pounding as she walked toward the microphone that faced the center of the U-shaped tables and the members of the school board.

"I read about you, the board of education, in the *Argus* in October, and I ... " she paused, " ... I disagreed with your decision to use styrofoam instead of paper trays." She had trouble saying that she disagreed with them right to their faces, but when she had written her speech, she had decided that that was truly how she felt. Saying it in front of this panel of grown-ups was even more difficult than she had anticipated when she first wrote the speech, but she took a deep breath and said it anyway. She relaxed somewhat when there was little response from the row of elected officials who sat intently watching her. Unlike Mr. Stehle who had been disappointed with the lack of response from his class when he first told them about Tanja's current events assignment and the possibility of challenging the school board, Tanja gladly accepted the board members' blank stares as encouragement to finish up her presentation.

Even though she had practiced and knew exactly what she planned to say, she was still nervous. This prevented her from ad-libbing, and she read word for word from her notes, describing her current events article, the subsequent survey, the recently completed Trial Week, and its landslide victory in favor of paper trays. To support her stance, Tanja distributed copies of the survey and the "Common Sense" anti-styrofoam pamphlet to each of the board members.

" ... and since the students have proven that we are willing to pay extra money for paper trays, we think you should switch," Tanja concluded. But there was no applause, and the brief pause was uncomfortably silent. Finally, the school board president said, "Thank you." As Tanja sat down, Mr. Stehle whispered, "Very good" and nodded his head in approval and praise.

Then each of her classmates took a turn speaking, agreeing that paper trays are better for the environment and that the West Milford school system should go back to using them. Each student presented a list of facts in support.

Then it was the board's turn. They had listened attentively and were genuinely sincere when they asked the kids a number of questions about the use of paper vs. styrofoam trays. Even though the students, Tanja especially, had become knowledgeable about styrofoam's effects on the environment, their insecurity made them pass most of their questions off to their teacher.

"They seem so critical and unsure," thought Tanja. "They're never going to vote for paper."

They, mostly Mr. Stehle, answered question after question, and finally a number of motions were suggested, considered and reworked. The board's discussion seemed to indicate that they were leaning in favor of making the switch, but there were some skeptics.

"What about our inventory of styrofoam trays?" asked one board member.

"How are we going to pay the extra costs?" asked another. And the additional costs were substantial—$36,000 a year for paper trays compared with $14,000 a year for styrofoam.

Finally, one motion was proposed and seconded. Board member Connie Nobis said, "I move that we use up the styrofoam trays and cups that we have already purchased and then switch to paper trays and cups, at which time the price of lunch will increase by five cents."

"Second," said board member Joseph Duffy.

The vote began as the board secretary called the role.

"Mrs. Nobis?"

"Yes"

"Mr. Duffy?"

"Yes."

"Mrs. Banis?"

"Yes."

"Mr. Ekenstierna?"

"Yes."

"Mr. Richko?"

"Yes."

It was unanimous. On December 20, 1988, the West Milford Board of Education voted to phase out styrofoam lunch trays and switch to paper. Tanja had won, and the headline that led her to achieve this goal could now be changed from "Despite threats to all life, board chooses styrofoam" to "Despite extra cost, board considers environment and chooses paper."

After the vote, the students congratulated each other with high fives all around. Then the board members praised them, too, saying that they were proud that their students proved that this was an important issue to them. The board president even encouraged the students to convince the township council to also get involved with the anti-styrofoam campaign.

The kids felt like heroes. The adults had really listened to them.

The local paper ran a story about the decision, complete with pictures of the students. Tanja's father bought as many copies as he could find. The proud father, however, was not one to praise excessively, and he commented, "If you ever make it into *The New York Times*, then you've really made it."

More newspapers throughout New Jersey picked up the story and made Tanja a state-wide hero. And four months later, her father's comment really came true when *The New York Times* published an article about Tanja's campaign against styrofoam.

The Environmental Club and KAP
from left to right: Kurt Schneid, Chris Frezza, Evonne Kemble,
Renate Vogt, Shari Ragan, Jane Ragan, John Misi, Carol Misi,
Tanja Vogt, and Norma Sullivan

Chapter 4

The anti-styrofoam campaign spreads

A lot happened in the four months between the school board meeting and the article that *The New York Times* eventually published about Tanja on April 28, 1989. Within four short months, Tanja had gained international recognition for a campaign that she thought had ended when she left the board of education meeting that night in December. But that was just the beginning.

◦◊ ◦◊ ◦◊

"What good is it if just one school switches from styrofoam?" asked Dave Corsaro soon after the school board's decision. Dave, who was only 12 years old, was one of the middle school students who had helped Tanja make her presentation to the school board. He raised the question during a discussion in Mrs. Sullivan's class.

"It's a great victory," he continued, "but what good does one school do when you're talking about the whole planet?"

"It's a good start," countered Mrs. Sullivan. "What do you think?" she asked the entire class.

They all seemed to agree with Dave that it didn't do much good.

Mrs. Sullivan was surprised by her students' sudden change in attitude. "Well, are we supposed to take this as a defeat? Did we just succeed in getting our board of education to listen to us, and that's the end of it? Do you think that it really wasn't a victory, because, let's face it, the atmosphere isn't going to be saved by one school district's switching."

"Let's write to everybody in the country," one student suggested.

"That's a little bit ambitious," said Mrs. Sullivan, immediately putting the brakes on a nearly impossible task.

Eventually, after discussing the possible ways they could inform other schools about the problems with styrofoam, Mrs. Sullivan's seventh graders came up with the idea of writing letters to schools throughout New Jersey encouraging them to switch from styrofoam to paper. To get the job done, the middle school kids formed a local chapter of a national group known as Kids Against Pollution.

In the high school, Tanja and her fellow enviro-kids had also formed their own Environmental Club, which met after school every Tuesday. At the first meeting, 60 students and adults attended. Soon after, the group was reduced to about 25 serious, active members, but its core included sophomores Tanja, Kurtiz Schneid and Chriss Frezza, senior Damian Cavalli, and junior Greg Tremarco.

Both the high school and middle school clubs plus many of Mr. Stehle's and Mrs. Sullivan's classes participated in the letter-writing campaign. Mrs. Sullivan had randomly selected 700 schools throughout the state, and she divided these among the interested students in both schools. Although each student wrote individual, personalized letters to one person at each school, they all followed the same basic format. They each briefly explained what had happened in West Milford and urged other schools to do the same.

Because costs were involved, (700 stamps do not grow on trees nor fall from a depleted ozone layer), Mr. Stehle and Mrs. Sullivan had to convince the school administration of the educational value of the endeavor. The two teachers stressed the campaign's historic value and the fact that it helped the students practice letter-writing skills. Mr. Stehle had already related it to the American Revolution, and Mrs. Sullivan had even tacked up a large map of New Jersey in the back of her classroom. Colored pins showed which towns had been sent letters and which ones had written back. The map proved to her principal that the project even helped improve her students' knowledge of New Jersey geography.

"Look at the letter I got," Tanja announced to the club one day. "It's from Brielle Elementary School, and it says, 'We read about you in the newspaper about getting rid of styrofoam trays. We would like to do the same thing in our school because we have the same problem. We hope you can help us since you are high school students and we are third graders.'" Tanja was thrilled that she and her classmates were generating interest in other schools. "Isn't that great?" she said. "Other schools want to do it, too, and they're starting so young."

Soon all of the enviro-kids started looking for their own letters. "Did I get any mail today?" they would ask the two teachers every day. Unfortunately, the positive responses were rare. Most schools didn't even use styrofoam in the first place. Of 77 schools that responded, most said they did not use plastic. But of the 29 that did use styrofoam, some said they would try to follow West Milford's lead.

Schools in two New Jersey towns—East Brunswick and Maplewood—and even in Suffern, New York, wrote back saying they wanted to ban plastic trays, and a school in Ramsey, New Jersey, said it would survey its students to determine if they wanted to switch.

The students' anti-styrofoam campaign was growing, and letters of interest were coming from all over. Tanja even started getting letters of praise from politicians. Only one month after the school board's unanimous vote, she received this letter:

> *January 31, 1989*
> *Dear Tanja:*
>
> *You are responsible for making people aware of the absurdity of our fetish with plastic products. I commend you and your fellow students for having the resolve to do what is right and then doing it.*
>
> *Unlike European countries, we have become too dependent upon plastic products. We only have one environment and if this is the legacy we are going to leave our children, we better all rethink our direction.*
>
> *Tanja, the ripple effect of your action would surprise you. Keep up your endeavor and count on my support.*
> *William J. Pascrell, Jr.*
> *Assemblyman, 35th District Paterson, NJ*

And he was right, the ripple effect was working. Still more schools expressed interest. New Jersey schools in Clifton,

Englewood, Lake Hopatcong, Ringwood, Hawthorne, Salem, Sparta, and even Stockton State College wrote of their interest. And the mail brought Tanja more political praise:

> *February 1, 1989*
> *Dear Ms. Vogt:*
>   *I would like to take this opportunity to commend you on initiating the crusade to ban non-recyclable plastic products in your community. Your school essay has sparked a local awareness campaign and has succeeded in making not only your community and school aware but also others throughout the state.*
>   *It is refreshing to see a young person such as yourself show such a genuine concern for preserving our environment. I hope that we can all work together to promote a more safe and suitable environment not only for ourselves but for future generations.*
> > *Marge Roukema*
> > *Member of Congress House of Representatives*

Politicians and average citizens alike wrote Tanja to express their support and appreciation, and although Tanja's father copied the letters to show at the office, he teased, "I won't be impressed until you get a letter from President Bush." That came about a year later. Unfortunately, the former president misspelled Tanja's name:

> *March 21, 1990*
> *Dear Tanya:*
>   *I was pleased to learn of your efforts to clean up the environment. You can be proud of the time and energy that you gave to this worthy cause, demonstrating how each individual can make a difference in his or her community.*
>   *One of America's greatest traditions is that of neighbor helping neighbor. You are a shining example of this tradition. The care and concern you've shown serve as an inspiration to those around you.*
>   *Keep up the good work, and best wishes for the future.*
> > *George Bush*
> > *The White House*
> > *Washington*

Unfortunately, though, all of the letters weren't full of praise, especially the ones from the plastics manufacturers. Dow Chemical Company wrote to inform Tanja that the company owned the brand name "Styrofoam" and that she should refer to the plastic by the generic term "polystyrene." No problem, as far as Tanja and Mr. Stehle were concerned; they simply switched from using the word "Styrofoam" to using the word "polystyrene."

Other manufacturers wrote (some even called) to dispute the claims against polystyrene. They said that they had already stopped using CFCs, the material that Tanja had learned was responsible for making a hole in the ozone layer. A group known as none other than the Polystyrene Packaging Council sent a booklet to Tanja which stated that the companies are now using a water-based compound, HCFC-22, which was determined to be 95 percent better than CFC.

To this, Mr. Stehle responded that there is still that five percent of damage. He also discovered that HCFC-22 causes its own environmental harm, such as low-level pollution, and that when polystyrene breaks down it releases poisonous gases called phenols.

Other letters came from West Milford's own former supplier of polystyrene cups, Dart Container Corp. of Mason, Mich. This company even sent a representative to a school board meeting, most likely in an effort to restore its lucrative contract with the school. "I am not disagreeing with you that there are problems with polystyrene," said the company's representative, Bill Wertz, who wore a jacket and tie with a V-neck sweater that softened his corporate image. "We are not ignoring the problems. As an industry, we are trying to find solutions to the problems associated with polystyrene."

Mrs. Sullivan had to respond: "I don't understand why we have to use these products at all. They are dangerous, and there are alternatives to polystyrene."

"We are looking into recycling as a solution," he countered, comparing food containers made of plastic to those made of paper, which he said cannot be recycled. "Paper food products are not recyclable because they have a coating on them to make them more functional. Recycling foam is more realistic than recycling paper food products."

But Mrs. Sullivan had an answer for that as well. She said that there are many schools that use washable plates and that one of her students was conducting a survey of schools in New Jersey to see how many use washable plates and silverware rather than disposable products at all.

The meeting ended with both the plastic manufacturer and the anti-polystyrene activist standing firm, neither accepting the other's position.

But this didn't stop West Milford's enviro-kids. Encouraged by their successes rather than discouraged by their few set-backs, the students decided to build on their achievements by undertaking a letter-writing campaign to local businesses known to use polystyrene. The letters were similar to the ones they had sent to schools throughout New Jersey, but because the businesses were in their own town, the students followed up with phone calls and visits to encourage them to switch.

"I'm Tanja Vogt from the West Milford High School," Tanja would say when she called local businesses that used polystyrene. After describing the negative environmental aspects of the plastic, she would continue, "We got the school to switch to paper, so now we're trying to get the businesses in West Milford to do the same. How would you feel about switching from styrofoam to paper."

Often the business people were nice and simply asked for information about how to switch, but just as often they were negative: "People want their coffee in Styrofoam in the morning, and they won't want it in paper." Delis especially resisted the switch, offering such excuses as "It's too hot to hold coffee in a paper cup" and "The wax coating melts."

Businesses were not the only target. Even the government of West Milford got involved with the anti-polystyrene campaign. The subject came up at a township council meeting less than a week after the board of education made their decision regarding plastic trays. "I think we should ban Styrofoam from being sold in the township," said councilman Edwin Aldrich. "If the students care about their safety and their environment, then we should care."

However, the township attorney, Robert Martin, warned that there could actually be some legal problems with banning polystyrene. Still, others on the council agreed with the students.

Council member Dawn Sharkey suggested the council send letters to township establishments suggesting they stop using polystyrene. "I am curious to see if McDonald's, delis, and other establishments would comply," she wondered aloud.

So, the council's first tentative steps toward banning polystyrene resulted in the decision to send letters to encourage all establishments in the township to discontinue using the plastic.

A month later, the council became even more decisive and adopted a resolution on January 18, 1989, that stopped the use of disposable plastic products in township agencies and departments and urged township businesses, especially fast-food establishments, to stop using and selling disposable plastic products. The ordinance also stated that any person or entity wishing to use township property would not be permitted to use disposable plastic products. Tanja had made a difference, not only in her school but in her town as well, and the campaign continued spreading.

And as it did, newspapers continued contacting Tanja, and the letter-writing campaigns and the township council's resolution only increased their interest. It seemed that the teenager was announcing to her family on a weekly basis that she was the subject of another newspaper article. And when the articles would appear, Mr. Vogt stocked up on extra copies. Tanja's fame was growing.

Tanja and Dr. Ruth

Chapter 5

McDonald's not easily swayed

Kids Against Pollution and the Environmental Club became West Milford's enviro-crusaders. They took on every business in town, writing letters, calling on the phone and making personal visits to encourage local restaurants, delicatessens, and supermarkets to switch from polystyrene to save the planet.

They made presentations to the local Lions' Club about the dangers of polystyrene and received a donation for their efforts. Dave Corsaro was so impressed with the response from the Lions' Club that he notified all the Lions' Clubs in New Jersey.

The state's top environmental official, Christopher J. Daggett, commissioner of the New Jersey Department of Environmental Protection, visited the high school to congratulate the students during a special assembly.

In their homes, the kids convinced their families to act more environmentally. Even Tanja's mother got into the act. Wherever she went, she would not accept polystyrene. When she went to the mall with her friends, she wouldn't buy anything from a fast-food restaurant that served in polystyrene. But even she suffered from her own peer pressure. As she continued searching the mall for a cup of coffee in a paper cup, her friends, who had already begun drinking from polystyrene, would cajole her to "Hurry up and sit down so we don't finish before you even get started."

At home, the enviro-kids didn't limit their cause to polystyrene. They convinced their parents to use disposables as little as possible, to recycle, to stop using aerosols because they also broke down the ozone layer, and to basically be as environmental as possible. The students recycled in school. They placed containers to collect paper, aluminum, and other recyclables in the classrooms and the hallways, and recycling became the "in" thing.

They even raised funds environmentally by collecting aluminum cans to sell for recycling. They gathered and crushed the cans and brought them to a company that bought aluminum by the ton. Unfortunately though, it took a lot of crushed soda cans to add up to a ton, and this generated only a small amount of money.

But Tanja and Greg Tremarco, another member of the Environmental Club, soon discovered a way to use the cans they collected to make far more money with far less work. "Don't crush those cans," Greg called to Tanja one day as he jumped out of the pickup they were using to deliver the cans they collected. "Just load them all into the truck."

"Why?" asked Tanja. "You know the recycling place won't take the cans unless they're crushed."

"We're not going to the recycling place. We're bringing these cans to a place where we'll get much more than the pennies per pound the recycling place pays, and we don't even have to crush them."

"Yeah, right," Tanja asked disbelievingly. "Where's that?"

"I'll show you," he said, but he wouldn't tell her where. They loaded up the truck and headed over the mountain toward Warwick, the next town over. Upon passing the sign that said "Welcome to Warwick, Please Drive Carefully" not only did they cross the border into another town, they also entered another state: New York. Unlike New Jersey, New York requires a deposit of five cents on every bottle and can.

Greg found a parking spot big enough for the beat-up but smooth-running old pickup and parked as close as possible to the entrance of Lloyd's, Warwick's major supermarket. Together they carried as many cans as they could into the store. Tanja was beginning to understand. "Here it is," he said, pointing to a device that resembled a soda machine just inside the entrance. But instead of exchanging money for chilled cans of soda, this machine accepted empty cans and dispensed money.

"You can't do that. These cans are from New Jersey," whispered Tanja, looking from side to side to make sure nobody was listening. And she was right, the machine was meant to accept only New York deposit cans, and supposedly it was designed to do just that. As the empty cans were placed in the machine, a device "read" specialized codes written on them to determine if they had been purchased in New York.

In fact, the machine did regurgitate some unacceptable New Jersey cans, but that was the exception rather than the rule. It accepted most of the "illegal" cans and exchanged them for cold, hard cash. For more than four hours, Tanja and Greg emptied the pickup truck, loaded cans into the machine and collected money. Their hands and clothes became filthy, and their feet stuck to the syrupy residue that dripped from the hundreds of empty cans. They left Lloyd's with an empty pickup and pockets full of money to help finance the activities of the Environmental Club.

Those activities consisted mostly of encouraging the entire town of West Milford to follow the lead of the board of education and eliminate the use of polystyrene. And their plan was working. Not only had the township council phased out polystyrene; many local businesses also followed suit.

But one day, the teenage activists encountered a business that wouldn't change its environmentally hazardous ways. It was McDonald's, with fast-food outlets all over the world serving in polystyrene containers. One of these outlets was located right in the center of West Milford, home of Tanja Vogt, the Environmental Club, and Kids Against Pollution.

As with every food-related business in town, the kids had sent a letter to McDonald's asking them to switch from polystyrene to paper or reusable containers. The response from McDonald's was that they give the people what they want, and the people want polystyrene because it keeps things hotter than paper for a longer period of time.

"How do they know that's what the people want?" Tanja asked one day at an Enviro-Club meeting. "Did they ask everybody?"

"Probably not," said Kurt. "I guess we'll have to prove it to them like we proved it to the school. Why don't we do the same thing we did with the board of education. First we'll do a survey, and if that doesn't get them to switch, then we'll 'put our money where our mouth is' and buy from McDonald's only if they serve without Styrofoam."

"Polystyrene," corrected a number of the club members who had learned that the brand name was "Styrofoam" and the generic term was "polystyrene."

"Right, right, right," said Kurt, smiling.

But not everyone agreed that they should do a survey first. After all, that was the first step they had taken with the school board, and it had only put them in a position to do another survey with their wallets. "Why waste time?" they thought. So they all agreed they would skip the survey step and begin immediately by casting their vote at the cash register. They would eat at McDonald's only if the food was served without polystyrene. Mr. Stehle, advisor to the Environmental Club, proudly watched his young activists continue their revolution.

Members of the Environmental Club, KAP, and their families and friends continued eating at McDonald's, but as they stood at the stainless steel counter and selected their Big Macs, McNuggets, or other "McFood" from the bright, backlit "McMenu," they always asked that their food be served without polystyrene.

At first, McDonald's accommodated their requests, always willing to give their customers what they wanted. But when more and more customers requested paper instead of plastic, it finally became a nuisance for the restaurant. It also started cutting into the restaurant's profits, because having alternative packaging available had become an added expense. Eventually, the fast-food restaurant became reluctant to keep switching back and forth from polystyrene to paper. The store changed its policy and offered only polystyrene. When people requested paper, they got polystyrene along with a booklet describing McDonald's reasons for using the hazardous plastic. Like Thomas Paine and West Milford's Environmental Club, McDonald's began circulating its own pamphlet.

"They won't even listen to us," Tanja said at an Environmental Club meeting after they discovered that McDonald's had reverted to using only polystyrene. "They won't even listen to their own customers. At least the school board listened to us and switched when we proved that we would really pay extra for paper. McDonald's won't even switch when their own customers ask for paper."

"Maybe they want a nickel more per customer, too," Kurt joked, and everyone laughed, recalling how it had been their money and not their words or surveys that had convinced the school board that the students truly cared about the environment.

"No, actually what they say is that they plan to recycle the polystyrene, or they might even burn it," Tanja reported to the club. "Right now they're conducting a pilot program in Brooklyn where they will take the styro-trash from nine McDonald's and recycle it. One of the things they hope to make from it is building insulation. They're also considering burning the stuff to get rid of it. Listen to what McDonald's says," she said as she pulled out a newspaper clipping. "This is about us, guys," she announced then began to read from the article, "McDonald's media relations representative, Linda Fontana, said, 'I'm not sure why the students are doing this. Tests on polystyrene show it's recyclable and, when incinerated, the only thing it emits is carbon and water vapor. It's a clean burn.' Endquote," said Tanja when she finished reading the portion of the article. "But according to someone Mr. Stehle called for information ... What was her name?" Tanja asked her teacher, the club's advisor, who sat on the window ledge observing the meeting.

"Ummm, Deborah Wallace. She's from the Center for Biology of Natural Systems in New York," he said, nodding his head as a means of confirming his own memory.

"Yeah," continued Tanja. "She said that using plastic as insulation is dangerous. And when it's burned it emits poisons called phenols. Burned improperly, it can even be cancerous."

That's when the Environmental Club decided they had to be heard. If McDonald's was going to recycle, they would "help" them. They would bring back so much polystyrene that McDonald's wouldn't know what to do with it. After all, they argued, McDonald's was primarily a "take-out" restaurant, and any polystyrene that was taken out would not be recycled. They would "help" them, and by returning their "styro-trash" to the restaurant they hoped to convince McDonald's that switching to paper was really the more environmental solution. So they collected the polystyrene littering the country roadsides throughout West Milford and even rescued the styrotrash from garbage cans before it could be sent to the overflowing landfills.

While some of the slimy garbage was destined for the McDonald's restaurant in West Milford, some of it would also be sent to Joan Kroc, the surviving wife of the founder of McDonald's and the international corporation's chief shareholder. She had been an antiwar activist during the Vietnam War, and the young environmentalists thought she might sympathize with them. If not, at least she might encourage her company to do something to stop the avalanche of slimy polystyrene trash coming to her San Diego residence via the U.S. Postal Service.

And it would become an avalanche because the West Milford students were not the only ones sending styro-trash to Joan Kroc. The idea for the "Send-It-Back Campaign" originally came from Lois Gibbs, the director of an environmental group known as Citizens' Clearinghouse for Hazardous Waste in Arlington, VA. She had experience convincing major corporate powers to change their polluting ways, and, like Tanja, she was prepared to fight McDonald's.

Lois Gibbs had already been helping promote the anti-polystyrene campaign and had been using Tanja and her success with the board of education as an example to encourage other schools to do the same. Now they would work together on a much larger and more powerful entity.

꙳   ꙳   ꙳

The "Send-It-Back Campaign" became a nationwide program of activists sending their styro-trash to Joan Kroc. The West Milford enviro-kids wanted to be heard and wouldn't rest until they convinced McDonald's to switch.

Posters promoting Operation Send-It-Back were put up throughout the high school and middle school. They read:

Send your Styro-Trash back to McDonald's because:

1. McDonald's creates 1.6 billion cubic feet of styro-trash a year
2. Styrofoam is useful for minutes but is here forever
3. Chemicals used in styrofoam can end up in your food
4. Production harms workers
5. Chemicals used to make styrofoam add to atmospheric pollution
6. Burning styrofoam releases toxic gases
7. Safe alternatives exist! (Burger King, Wendy's, Roy Rogers use paper)

The students gathered styro-trash and packaged it for its cross country trip to the doorstep of McDonald's chief shareholder. But while sending the dirty plastic packaging back to Joan Kroc was noticed primarily by the company itself, sending styro-trash back to the McDonald's restaurant in West Milford became a public event observed not only by the company but by the township's residents, the media, and the millions of people reached by the newspapers, television, and radio stations that took notice.

When KAP and the Environmental Club brought bags of polystyrene back to their neighborhood McDonald's, it was like a Fourth of July parade or an important local sporting event. Photographers, reporters, and even television cameras were there to record the events for the media, the police observed from a safe distance in case they were needed, and some spectators simply watched while others joined in.

The first "Send-It-Back" rally occurred in April of 1989, only four months after Tanja and her classmates had first convinced the school board to switch from polystyrene to paper trays. A group of about 50 students, some teachers, parents, and a few interested township residents gathered in the McDonald's parking lot on April 27, 1989. Mr. Stehle had warned the owner and manager of McDonald's that they were coming, Tanja was prepared with a speech that she planned to read to the owner, and Kurt sat in a shopping cart borrowed from the nearby ShopRite. Sitting in the cart, Kurt was buried under a pile of used styro-trash—coffee cups, hamburger clamshells and Egg McMuffin trays—all destined to be returned directly to the owner of this particular McDonald's franchise. The trash had been taken from McDonald's own garbage cans and from along West Milford's streets.

Kurt, however, was no longer Kurt. Sitting in the shopping cart he had transformed himself into Ronald McToxic, a clown with a red frown painted on his face, the rest of which had been painted white. He wore a baggy clown suit, oversized clown shoes that hung over the back of the shopping cart, and a purple afro wig. Ronald McToxic was sad and would remain sad until McDonald's stopped using polystyrene in its restaurants.

He represented the harm that polystyrene causes to the environment, and he wore a sign that said, "McDonald's uses: 1) non-biodegradable toxic polystyrenes 2) chemicals which add to atmospheric pollution 3) 1-6 billion cubic feet of styro-trash per year."

Kurt wasn't the only McClown to debut that day. To balance Ronald McToxic's negativity, Ronald McFuture was on hand as well. Although he had a boy's name, Ronald McFuture was actually played by a girl, Chriss Frezza, another member of the Environmental Club. She represented the positive results that could occur if McDonald's switched from polystyrene to paper. The sign she wore called for:

> **Food waste/paper composting**
> **Reusable travel jugs and travel bags**
> **Paper packaging**

Other students carried signs that read "WMHS Styrofoam Free," "Catch Recycling Fever," and "Be a Good Neighbor to Our Planet."

The crowd marched across the parking lot led by Tanja and Mr. Stehle while 17-year-old junior Todd Case pushed Ronald McToxic in the shopping cart filled with styro-trash.

Tanja made her case, reading from a speech that detailed the dangers of polystyrene. Ronald McToxic and Ronald McFuture also spoke, presenting both the harm caused by the plastic and the good that could result from switching to paper. Other members of the Environmental Club and KAP also presented their sides to the restaurant's representatives. They warned them that they planned to make monthly deliveries of the styro-trash to "help" them recycle.

William Pascrell, the Democratic assemblyman from Paterson, West Milford's nearest city (the man who wrote to Tanja about the ripple effect she would cause) also came to support the enviro-kids. He announced at the rally that he would introduce legislation the following month requiring New Jersey to recycle 50 percent of its residential garbage by 1994, up from the existing law requiring 25 percent recycling. The crowd cheered.

Additional support came from local environmentalist Doris Aaronson, Environmental Trustee of Pinecliff Lake, a lake community in West Milford. She called for more biodegradable substitutes and said that when Pinecliff Lake had its drawdown, over 100 plastic items were exposed.

But the franchise owner, George Slider, Sr., and his son, the manager, George Slider, Jr., were just as prepared as the enviro-kids. The senior Slider maintained that polystyrene is what the people want because it keeps "hot food hot and cold food cold." He sounded like a commercial. "And even though polystyrene really isn't even a solid waste problem (it only amounts to one-quarter of one percent of the material found in landfills), McDonald's is already working on recycling and incineration programs to remove the material from the waste stream," he added.

"What about the poisons that are emitted when it's burned?" asked Tanja, pleased to be able to debate him with some facts of her own. "What about the phenols that come from burning styrofoam? The poisons."

"As far as I know," began Mr. Slider, looking to his son for support. "As far as I know, polystyrene emits only carbon and water when it's incinerated."

No matter what Tanja and the other environmentalists said, the restaurant owner maintained his belief that polystyrene was harmless and better than paper.

Finally, Mr. Stehle convinced him to schedule a meeting in May where both factions could present their sides. Mr. Slider concluded, "We'll handle this in a responsible manner, and we'll get back to you."

By now, Ronald McToxic had climbed out of the shopping cart, and Todd wheeled it right under Mr. Slider's nose, delivering the smelly cart of styro-trash. The enviro-kids promised to make a delivery every month.

As the students boarded the bus to leave, George Slider pushed the cart behind the restaurant near the trash bin. When a newspaper reporter asked him what he intended to do with it, he was a little perplexed about the gift of trash. "What will I do with it? Well, I guess I'll discard it in my compactor. I don't know what I can do with it. Do you have any suggestions?"

Chapter 6

Tanja addresses the United Nations

At about the same time that Ronald McToxic and Ronald McFuture first confronted the owner and manager of their local McDonald's, some of the members of KAP and the Environmental Club were invited to a different kind of meeting that promised to be less confrontational and more international.

"They want me to speak at the United Nations," Tanja announced to her family one night at dinner.

"What?" exclaimed her mother, both surprised and proud. She practically jumped out of her seat.

"What do you mean, speak?" asked her father. "They must mean listen to speakers."

"No. They mean speak," continued Tanja. "Mr. Stehle told me today. They're having some kind of meeting of young environmentalists. They've been reading all of the articles about me, and they invited some of the kids from the Environmental Club and Kids Against Pollution to speak at it."

Mrs. Stehle gave her a big "I'm-proud-of-you" smile that looked like she should be pinching Tanja's cheek as well. "That's great. I can't believe how far this thing is going."

◆◇ ◆◇ ◆◇

To accompany 15-year-old Tanja to the U.N.'s first International Youth Environmental Forum on May 1, 1989, were Kurt, 16, Greg Tremarco, 16, John Nisi, 12, and Sheri Ragan, 14. As the big day came closer and closer, Tanja grew more and more nervous about her impending speaking engagement.

When the day finally arrived, Tanja stood before a pile of clothes thrown carelessly on her bed. "No, I can't wear this. I'd look stupid," Tanja said throwing another dress on top of the pile. She had been through so many outfits that most of her clothes were there instead of in her closet. Tanja was frantic. She was so nervous, she was driving her whole family crazy. Downstairs, her mother continued ironing more potential outfits for her.

Finally, not long after she had settled on something to wear, Tanja was headed toward New York aboard a school bus full of rowdy kids. Her parents drove separately to the U.N. For the lucky students attending the U.N.'s Youth Environmental Forum as spectators and not speakers, this was simply another field trip, another opportunity to get out of school and ride a bus somewhere, anywhere away from the school building where they were usually sequestered. While everyone else talked and screamed, Tanja sat alone on the bus, trying to concentrate on the notes in her lap, making some final touches on the speech she had spent the last week preparing at home.

Once the bus arrived, Mr. Stehle and Mrs. Sullivan whisked Tanja away from her classmates on the bus and shuffled her into her seat in the U.N. general assembly, where they left her to make her final preparations. The room was abuzz with activity. Television camera crews were setting up, amplifiers were being tested—"Test, 1, 2, 3, test"—and the panel of international environmentalists Tanja was to address were taking their seats.

While a large group of young environmentalists from throughout the world had been gathered together to speak, Tanja was among the first, and with a CBS camera and its red "filming" light only inches from her face, Tanja began her speech.

As she spoke into her microphone, she tried to ignore the news camera, but it made her even more nervous. Although she had intended to use her notes only as an outline, her ad-libbed lines would not flow smoothly, and she was forced to rely less and less on her memory and more and more on what she had written. Eventually, she resorted to reading her speech into the microphone.

She told the story of how she and her classmates eliminated polystyrene from their school, how they were trying to help other schools do the same and how they were also trying to get polystyrene out of their town. At the end of her story, she concluded, "Please don't let anyone tell you that one person can't make a difference," and that was the quote that made the CBS newscast that evening.

It was her first time in front of a camera but hardly her last. In fact, that same day she appeared on two television broadcasts and was interviewed by two radio stations. Immediately after the U.N. speech, a radio station employee whisked her by cab to the WNSR studio, where she was interviewed about her cafeteria tray campaign and where she basically reiterated the information from her speech. Without a camera and its red light shoved in her face and encouraged by the success of her presentation at the U.N, Tanja was less nervous for the radio interview. She referred to her notes only occasionally as she repeated the story that would soon become world renowned.

After that, she was driven back to the U.N. where she joined Mr. Stehle for another radio interview and repeated her story. Her nervousness faded away more and more each time she retold the tale.

One of the few breaks she had that day was for lunch at the U.N., where she joined the other speakers who were also taking a break from the hours of speeches being presented that day. But unlike the television and radio interviews, it was Kurt and not Tanja who was the star of the lunchroom. During his U.N. speech, he had described how he had dressed as Ronald McToxic the previous week to return styro-trash to McDonald's. "We intend to return to McDonald's on a regular basis, and we invite all of you to join us!" he had announced and concluded, "Maybe then Ronald McFuture will have a chance."

Now, in the U.N. cafeteria, the crowd cheered when the frowning clown suddenly appeared. Kurt worked the room as Ronald McToxic, generating more support for the anti-polystyrene campaign.

Afterwards, it was Tanja's turn to reenter the spotlight. That night, she and Mr. Stehle were again interviewed, this time by New Jersey's public television station, NJN. By this interview Tanja had repeated her story so many times that she no longer needed her notes. She had become an instant enviro-celebrity, and in just one day she had become an experienced speaker and interview subject for radio and TV.

It was around the same time that Mr. Vogt's joke had come true and the article about Tanja and her campaign appeared in *The New York Times*. With the attention brought by the article, the U.N. appearance, and the television and radio interviews, word of Tanja's campaign quickly spread.

Chapter 7

An education in plastics

Back in West Milford, the students continued their anti-polystyrene campaign and concentrated on McDonald's, their most prominent adversary. They continued sending garbage to McDonald's chief shareholder, and even though it might not have convinced the corporation to cease using polystyrene, it did get the attention of the executives there. In fact, they offered to fly Tanja and Mr. Stehle to their headquarters in Oakbrook, Ill., all expenses paid. They wanted to give the environmentalists a tour to educate them about McDonald's activities, particularly its use of polystyrene.

"What do we want to see them for with all their garbage?" Tanja asked Mr. Stehle. "Let them come here."

And they did. Four executives from the McDonald's headquarters flew in to address the West Milford students about polystyrene. Members of KAP and the Environmental Club plus a group of other interested students and teachers were there to hear McDonald's official response to the students' concerns. But they were in for a surprise.

The executives handed out plastic pellets and used a display board to present all of their facts about polystyrene. They described the manufacturing process and then claimed, "It's recyclable." They explained how recycled styrofoam can be made into park benches and other hard plastic products, and they said their goal was to recycle 25 percent by 1995.

"Not enough and too late," thought some of the skeptical students while others wondered, "How many plastic park benches do we need?" Mr. Stehle and Mrs. Sullivan already knew that after polystyrene is recycled once it cannot be recycled again.

But the audience was polite and waited for their turn without interrupting the speakers.

Unfortunately, the executives didn't stay long. As soon as they had made their presentation, they started packing up their display boards and plastic pellets to leave. They never even addressed the students' complaints about McDonald's continued use of polystyrene and didn't even invite the audience to ask questions or discuss their side of the issue. They hadn't once mentioned the damage polystyrene causes to the earth's protective ozone layer.

Nervous and unsure of whether it was their turn to speak, the kids were as silent as they had been the day in Mr. Stehle's class when he had asked them to start an environmental revolution. As the four McDonald's executives packed up and walked out the door, the students sat in shock that their complaints had been completely ignored.

But Mr. Stehle and Mrs. Sullivan were worried. It seemed that some of the kids might swallow the information they had been fed. As they walked back from the high school to the middle school, KAP member Laura Moskin asked Mrs. Sullivan, "Well, what's wrong with it if it can be recycled?"

Mrs. Sullivan said, "It is a good feature. I just wonder how many people are going to take their hamburger clamshells back."

At least she knew of some people who *were* taking their clamshells back. In fact, they were taking *everyone's* clamshells back, plus coffee cups and any other polystyrene packaging generated by McDonald's. KAP and the Environmental Club continued their Send-It-Back Campaign and held a rally each month in the McDonald's parking lot. Tanja, Ronald McToxic, Ronald McFuture, and the other young environmentalists brought polystyrene garbage and anti-polystyrene messages to McDonald's owners, managers, and corporate executives. The campaign spread, the rallies grew and more and more people learned about the environmental damage caused by polystyrene.

Finally, someone on the side of polystyrene agreed to address their concerns, not just give them a one-sided education in plastic like the McDonald's executives had. Since McDonald's plastic packaging was manufactured by Amoco Foam Products Co., a subsidiary of oil giant Amoco Corp., it was agreed that someone from that company would debate the pros and cons of polystyrene. Now it was up to KAP and the Environmental Club to produce someone with the ability to debate a well-informed executive from a plastics manufacturing corporation. And the young environmentalists did their job well.

With money they had raised through a variety of environmental fund-raisers, such as collecting aluminum cans, KAP and the Environmental Club flew Pam Peck from Vermont to New Jersey. Pam worked with a group known as Vermonters Organized for Cleanup, and she had received a financial grant that allowed her to support herself as a full-time environmentalist. The students had heard about her through Lois Gibbs' Citizens' Clearinghouse. She arrived a few days before the debate, explained her tactics to the students, and predicted how she expected her opponent to try to confuse the issue.

When Amoco learned that the press would cover the debate, the huge oil company tried to back out. "If you back out, I will go right to the press and scream," Mr. Stehle told them. Whether they were there or not, it would be reported in the news, so Amoco sent James C. McLellan, director of waste management and chemicals, to defend their side.

The high school auditorium was crowded on the evening of May 17, 1989. There were over 300 students and interested adults there that night. By now, the entire town of West Milford plus others throughout the state and in neighboring states were aware of the dispute. They had turned out to get a clear understanding of the facts on both sides of the issue.

Pam and Mr. McLellan stood behind podiums at opposite sides of the stage. Pam didn't look much older than a high school student herself. With her wide glasses and long, straight hair, she could have passed for one of Mr. Stehle's or even one of Mrs. Sullivan's students. Mr. McLellan, also wearing glasses, presented a corporate appearance with his short hair, white shirt and tie but softened it by rolling up his sleeves to give him that hands-on look. His moustache also made him look more casual.

During the debate, Pam clearly expressed her side. "Continue to boycott this dangerous product," she urged the members of the audience, supporting the anti-polystyrene campaign. "Eliminating this product is what we want. The product is dangerous all around."

But Mr. McLellan said that many people who were calling for the ban do not know all the facts about the product. He said those in the plastics industry do not believe the general public wants to ban the use of polystyrene products.

But Pam disagreed and supported her beliefs with facts: "More than 50 schools and other public organizations in Vermont have boycotted plastic foam products because of the health and environmental implications of polystyrene. The toxic chemicals from polystyrene go into the air, land, and water."

Mr. McLellan argued that the students' prime complaint is that polystyrene damages the ozone layer. To refute this, he said, "Manufacturers of polystyrene food containers now use HCFC-22 instead of chlorofluorocarbon in the manufacture of polystyrene, and HCFC-22 is 95 percent less damaging to the ozone layer."

But Pam found a flaw in his argument: "Although HCFC-22 is 95 percent less damaging than CFC, it is still hazardous. Plus, pentane, used in the production of foam products, causes emissions of substances that are the prime ingredients of smog."

She cited other harmful effects as well: "A recent fat tissue study found that 100 percent of all fat samples taken from United States residents contained styrene. The group conducting the study listed polystyrene cups as the probable source of the styrene."

"But the national toxicology program hasn't included styrene on its list of toxic chemicals," countered Mr. McLellan, "and any possible health hazards regarding possible effects to the nervous system and purification system are temporary. Packaging engineers themselves even recommend polystyrene for use in packaging food." He then went on the attack and responded to the environmentalists' complaints that polystyrene is not biodegradable. "Paper food containers are not biodegradable either," he said, "so both paper and polystyrene have to go to either composts or landfills. And polystyrene is actually better because we are working on ways to recycle it. We can even incinerate it, and the only by-products of burnt polystyrene are water and carbon dioxide."

Pam disagreed and cited studies that found that 57 chemicals are produced when polystyrene is burned.

"The study identifying 57 chemicals is misleading," said Mr. McLellan. "The number of chemicals released depends on the conditions during incineration."

After the debate, which lasted for over an hour, high school sophomore Shane Gosdis observed what many of the others in the audience would agree with, that Mr. McLellan had not given one good reason for people to continue using polystyrene instead of paper.

Soon after the debate, after its details and outcome were published in many newspapers, more oil companies and plastics manufacturers came down hard on the enviro-kids of West Milford. They called and sent letters describing what they would do in response to the polystyrene problem, how they would recycle, and how there would be a loss of jobs if polystyrene manufacturing was stopped.

One day, Mr. Stehle fielded a call from a plastics company. To support his side, the corporate representative told Mr. Stehle that too many jobs would be lost if they stopped making polystyrene. Mr. Stehle compared it to the invention of the automobile and the resulting demise of the horse-drawn wagon. "You mean to tell me that all those people who used to make wagon wheels were out of work forever because Henry Ford invented the car?" he asked rhetorically. "No, actually more jobs were created but just in a different industry." There was silence at the other end of the line.

But Mr. Stehle still was worried. He could see how the same kids who had been so successful with the school board and town council were starting to become fearful of these companies. The companies seemed to know more than they did, and the kids started losing belief in themselves.

◆   ◆   ◆

"And we thought the board of education was hard to convince," Kurt said at one of the Environmental Club meetings.

"Yeah," agreed Tanja. "They were nothing compared to McDonald's, and now they've got the oil companies on their side. How will we win? Even the United States government can't fight the oil companies."

"No, they work for them instead," joked Kurt, with a note of seriousness in his voice.

"But they just won't listen to us. They don't care about the ozone layer, the low-level smog, the litter or the landfills," complained Tanja, listing off the top of her head the harmful effects of polystyrene. "Can't they see that they're poisoning the earth?"

"Of course not, all they care about is making money," answered Chriss.

"I told you," added Kurt, "they want each customer to pay a nickel more."

Everyone laughed, but Tanja saw some truth in his joke.

"You're probably right," she said. "They probably haven't felt it financially yet, so all they're trying to do is get us to shut up before we do start making them lose money. We've got to prove to them that their customers really care. We've got to prove to them that their customers don't really want polystyrene."

"But all they keep telling us is that they're only doing what their customers want, and their customers want polystyrene," said Kurt.

"Yeah, but why do they really think that?" said Tanja.

"Because it keeps the hot food hot and the cold food cold," a number of club members said in unison. They had heard it so many times before that it had become embedded in their brains.

"Now we all sound like a commercial," groaned Tanja. "No, really. What can we do to convince McDonald's that their customers really want paper instead of plastic."

"I guess we have to do the survey after all," said Kurt. "We've got to find out what the people want and inform McDonald's of the results."

So, although they had tried to bypass the survey because of the experience they had had with the school board, the enviro-kids all agreed that a survey was now the next logical step toward convincing McDonald's to switch from polystyrene. But first they had to get their facts straight.

To determine if polystyrene really did "keep hot food hot and cold food cold," the members of KAP performed some tests in their middle school science classes using hot liquids, polystyrene cups, and paper cups. Their results proved that after six minutes polystyrene kept coffee or tea only one-half a degree hotter than paper.

With this information, they were ready to conduct their survey. It read:

*We, the Environmental Club
at West Milford Township High School,
in an attempt to come up with
a plastics pollution solution,
want to know—*

IS THIS WHAT THE PEOPLE WANT?

McDonald's says: The people want their coffee/tea to be 180°. That's why we use polystyrene cups.

The Opposition says: After 6 minutes, boiling water in a paper cup was a half a degree cooler than boiling water in a polystyrene cup.

**Q. What do you want? _____Polystyrene _____Paper
_____Reusable cups**
A. McDonald's says: The people want their food fresh & hot. That's why we use polystyrene.
A. The opposition says: Burger King, Wendy's, & Roy Rogers serve fresh & hot food wrapped in paper.

Q. What do you want? _____Polystyrene _____Paper
If you are old enough to read this,
you are old enough to answer:
Name_____ Address_____ Age_____ (optional)

Return to: West Milford Township High School
Environmental Club
67 Highlander Dr.
West Milford, NJ 07480

KAP and the Environmental Club distributed the questionnaire throughout the town, and Mr. Stehle got it published in all the local papers. Again, the vote was a landslide in favor of paper—90 percent.

"Now, what do we do?" the kids asked at their next meeting, and they decided it was time to present the facts to McDonald's.

It was now June, more than six months since the school board had decided to switch from plastic to paper. Summer was coming and the school year was nearing its end. On Saturday, June 10, 1989, KAP and the Environmental Club scheduled another rally at McDonald's to present the results of their survey.

The rally was bigger than any held before. Representatives of Clean Ocean Action, a coalition of environmental groups in the Highlands section of New Jersey, returned a whopping 189 bags of styro-trash. The garbage had been collected by high school students along a rare protected section of New Jersey shore at Sandy Hook National Park. Parkland or not, the cleanup proved that Sandy Hook was as susceptible to litter and pollution as anywhere else in the state, or the country, or the world.

Everyone was getting into the act. Also in attendance were fifth grade students from the Halsted Street School in Newton, N.J., a town that was even more rural than West Milford.

At the rally, Tanja read the survey results to George Slider, owner of the West Milford McDonald's franchise: "Of 2,000 people who responded to our survey, 1,409 people chose paper cups, 456 chose reusable cups and 76 people chose polystyrene. Regarding the clamshell package, 1,839 people preferred paper wrapping, and 82 preferred polystyrene."

But the students' efforts remained unsuccessful. Even after the survey, McDonald's maintained its pro-plastic stance. The response came from McDonald's vice president, Shelby Yastrow, "I know about surveys. I know how easy it is to get the answers you want."

Chapter 8

Disorder in the court

Even though McDonald's wouldn't listen, the teenage enviro-kids continued speaking and spread their word wherever it would do any good. Up until this point, West Milford had only adopted an anti-polystyrene policy for township offices and agencies themselves but had not gone as far as making a law that would ban the use of the polluting plastic within its borders. However, the Environmental Club did learn that the city of Paterson, an inner city located in the same New Jersey county as rural West Milford, was considering just such a law. Even before that, the council of Paterson was also considering implementing its own recycling ordinances, something West Milford had had on its books for some time. For Paterson, it was already seven months past the state deadline for adopting a recycling ordinance. The student environmentalists made sure they were there to offer their own educated guidance when the Paterson council made its decision regarding recycling and the use of polystyrene.

Although it was during the summer, August 1989, and school was not in session, Tanja, Mr. Stehle, and Mrs. Sullivan were able to gather enough club members to make a united appearance at the council meeting. And although they weren't dressed as Ronald McToxic or Ronald McFuture, Kurt and Chriss were there as well. They had all become as familiar with council meetings as Tanja had become comfortable at being interviewed, and they waited patiently for the official routines that got the meeting underway.

"I'd like a description of the proposed ordinance," insisted councilman Thomas C. Rooney, appearing belligerent as soon as discussion about the new law began. So Paterson's recycling coordinator, Nadine Mariconda, described the law to him, explaining how residents would be required to separate recyclables such as bottles and cans from the trash. She described the economic benefits to the town and the potential penalties for residents who did not participate.

"Do you really think residents will clean bottles and cans?" asked Mr. Rooney, who didn't even wait for an answer as he continued to list his reasons for opposing the law. "Why are the penalties so high—$1,000 fine and up to 90 days in jail? And how will we inform Paterson's residents about this new law? Many of them are old and can't see well, and many are foreign and don't read English." Then he started yelling. "I'm tired of generalities. I want specifics. I don't want to hear about Never-Never-Land again!" he screamed at the shocked recycling coordinator. The audience of residents plus the visitors from West Milford were shocked as well.

Bang! Bang! Bang! Council president Albert P. Rowe hammered his gavel to quiet Rooney. Then Ms. Mariconda explained that she would educate the young Patersonians through school and their parents and other adults through newspaper and television ads.

Then the audience was offered the opportunity to speak. First up was new resident Adele McCollum who said, "I can't believe the people of Paterson are as stupid as Rooney thinks."

But then Rooney, the aggressive councilman, verbally attacked Ms. McCollum and people like her "who move into housing that is tax-abated and then say bad things about Paterson without really knowing the city." Ms. McCollum tried to respond but was shouted down by Rooney who just got louder and louder.

Again, council president Rowe banged his gavel many times, but the shouting continued. Some observers laughed. The council was completely out of control, and council president Rowe got up and walked out, adjourning the meeting as he did. He hadn't taken an official vote for adjournment, however, so now some other council members joined the yelling. Others simply looked embarrassed.

Eventually order was restored, and the audience again began debating the new ordinance, each speaker taking their own turn at the microphone to address the council. After about ten minutes, Mrs. Sullivan took her turn. She introduced herself from West Milford and expressed her support for the recycling law. She said that people in her town don't object to washing bottles and cans.

Then it was Mr. Stehle's turn, but as soon as he said that he was also from West Milford, Rooney burst out yelling again. "What are you doing here?" he screamed. "We don't go to West Milford and tell you how to live, do we? This is Paterson. How many homeless do you have? How many drug addicts?"

But the council president hammered his gavel to maintain order, and Mr. Stehle was able to finish offering his support for the new recycling ordinance. Then it was Chriss's turn to address the council. She timidly walked up to the microphone, glancing fearfully up at Rooney. She was afraid of the verbal thrashing he might give her. Mr. Stehle and Mrs. Sullivan had to encourage her, saying, "It's alright. Just say what you feel."

In response to the councilman's concern that some residents don't speak English, Chriss suggested that they print fliers in Spanish and Italian as well. Even that didn't meet with Rooney's approval.

"What about the Palestinians?" he said. "You know if we went to West Milford like these people came here, they'd run us out of town." Again, he got louder and louder, and Chriss slipped away from the microphone. Rooney continued his ranting, "I've never seen anything so offensive in twenty years of public life, this phalanx from West Milford."

To this, a quick-witted member of the audience responded, "Take a look in the mirror when you're shaving tomorrow for something real offensive."

For hours the meeting continued in all of its disorder and disagreements, but by the end of the evening the council adopted the new recycling ordinance. After the meeting a number of the other council members apologized to the visitors from West Milford.

The council members were not the only ones to apologize. Although he didn't attend the meeting, even the mayor of Paterson himself sent an apology:

To: Ms. Norma Sullivan and Ms. Chriss Frezza
From: Frank X. Graves, Jr. Mayor
Date: August 31, 1989

It has been brought to my attention that at a recent meeting of the City Council of Paterson, some who had taken their time in the interest of the environment were severely chastised and ridiculed by members of the Governing Body. The Governing Body is an independent part and not under the jurisdiction of the Chief Executive. Nonetheless, please permit me as the Chief Executive of this City to apologize for the apparent rude treatment extended.

I am delighted that people who care took an avenue to pursue what they feel regardless of whether or not they live here. I extend to you the hope that you will not judge the entire City by the treatment afforded you.
I am not apologizing for any member of the Council, but as the Mayor of this City.

And it wasn't long before the enviro-kids again appeared before Rooney and the other Paterson council members, this time to support a city-wide ban on polystyrene. It was September 12, 1989, only a couple of weeks after the previous meeting, when the students came to try to make polystyrene illegal in Paterson.

While they had come to support the ban, their old familiar rivals were there to oppose it—McDonald's—both a Paterson franchise owner and a corporate executive. Still, the Environmental Club did their best to present the facts.

Tanja reminded the council that with McDonald's involved, it was more than just a local issue in West Milford. And she told them, "We surveyed the town and found 95 percent of the people we spoke with favored the environment over plastic."

Kurt then reminded the council that similar polystyrene bans had already been adopted in other municipalities throughout the country, including Berkeley, Calif., Vermont, Suffolk County, N.Y., and several Connecticut towns.

The McDonald's representative in attendance, Judy Donaldson, reiterated the corporation's usual response, "Paper can't be recycled, but polystyrene can." She said that the used plastic can be made into cafeteria trays, insulation, and even playground equipment.

Again, councilman Rooney opposed the young environ-mentalists. "It's easy for you to suggest closing down McDonald's," he said, "but the McDonald's restaurants in town are important sources of employment for Paterson residents. Everyone talks about clean air, but I suspect you came here by car, which spewed chemicals into the air. Would you suggest banning the automobile because it kills 100 people a week? Should we ban airplanes? Should we go back to living in caves?"

The Paterson council rejected the polystyrene ban by a vote of 6-3.

◦◇ ◦◇ ◦◇

As hard as McDonald's fought to support its continued use of polystyrene, Kids Against Pollution and the Environmental Club fought to eliminate its use. By November 1989, a full year after Tanja first read the "Despite threat to all life" article, the enviro-kids had held a number of monthly "Send-It-Back" rallies at McDonald's. By this time they were even willing to consider the possibility of recycling polystyrene, if it was done on their terms. And those terms were spelled out in a pledge that Tanja presented on the day of the November rally.

The air was cold but still as the kids gathered around the West Milford McDonald's. Mr. Stehle had notified the manager and the local press that the kids were planning another rally at the fast-food restaurant, and they were all there, notepads and cameras ready. Even the local cable access channel was there. Two cops sat in a marked patrol car at a safe distance.

Then the clowns arrived. They drove up in an old pickup truck, the same one Tanja and Greg had filled up with sticky, dirty soda cans to raise money for the Environmental Club. Kurt was in his usual Ronald McToxic get-up, wearing the sign that listed all of the harmful effects from McDonald's use of styrofoam. Ronald McToxic gave the thumbs down sign with one hand. With the other, he carried a clear plastic bag full of used polystyrene. Chriss returned as Ronald McFuture.

The other enviro-kids gathered around as the two clowns got out of the back of the pickup. They carried picket signs with such slogans as "Keep your trash out of our waste stream," and "Be a good neighbor to our planet," and "Ronald—Recycle Now."

The enviro-kids took turns speaking—first Tanja, then Dave Corsaro from the middle school, then the two Ronalds, McToxic and McFuture. For the umpteenth time, they told the owner and McDonald's corporate representative, Judy Donaldson again, everything they knew about styrofoam's harmful effects on the environment and why they felt McDonald's should stop using the plastic.

Then Tanja read the pledge that she wanted McDonald's to sign asking that they honor their commitment to recycling polystyrene:

McDonald's pledges to:

1. provide containers for separating polystyrene in-house by the week of Earth Day, April 1990.

2. accept all polystyrene packaging with the purpose of recycling for no less than 5 years, including styrofoam from delis.

3. recycle 70 percent of the polystyrene packaging since this figure constitutes the in-house business, and

4. institute the reusable coffee mug and soda glass as other fast food places such as Pizza Hut and Dunkin Donuts are doing.

Tanja handed the pledge to Judy Donaldson, McDonald's district manager from Bloomfield, N.J. And even though they had asked for a higher percentage of recycling than McDonald's had ever offered and for other commitments that McDonald's had never even mentioned, Donaldson's response surprised them. "I believe the corporate executives will agree to three of the four points on the pledge," she said.

Then Tanja reinforced the Earth Day deadline. "They have five months to get their recycling efforts going," she said. "If the restaurants do not begin recycling 50 percent of their polystyrene by Earth Day in April, we're going to call for a boycott of all McDonald's products."

Chapter 9

Publicity brings fame

In pursuing her anti-polystyrene campaign, Tanja became an enviro-celebrity. She was sought after for interviews on radio and television and in national magazines. At one point, it seemed to her that every week, sometimes alone and sometimes with the other members of the Environmental Club and KAP, she was the subject of a new article or television show about her successes and continued attempts to rid the environment of the harmful plastic.

Articles about Tanja and the polystyrene cause appeared in newspapers throughout the country. She was featured in practically all the top national magazines, including *Time, Glamour, Seventeen, Sassy, Reader's Digest, Family Circle* and *Good Housekeeping*. The *Time* article even carried a color picture of the clowns Ronald McToxic and Ronald McFuture.

The story also appeared in a number of specialized magazines such as *Environmental Action, Animal Kingdom, Homeowner, School Board News, Science World, Junior Scholastic, Patagonia, A Day in the Life of the Earth* and *The Environment, Action, World* and *Caring People*. The entire country was reading about Tanja Vogt.

Her fame was the result of hard work, and as the publicity about her cause increased so did the effort required to maintain the momentum. The articles generated thousands of letters from all over the country and even from other countries throughout the world. Many wrote simply to praise Tanja and her campaign, but many others asked for advice about what they could do to help. Tanja personally answered every letter and stayed up past midnight many school nights to ensure that both her homework and her letters were completed in a timely fashion. She even devoted her weekends to answering the thousands of letters she received.

And the publicity continued. Tanja and West Milford's enviro-kids appeared on television shows throughout the country and the world. The story was broadcast on CBS, on CNBC, on the Financial News Network, on ABC's *Good Morning America* and *Day's End*, and on PBS's *3-2-1 Contact* and *The Eleventh Hour*. Her story even made it to England on BBC, the British Broadcasting Corporation. Radio interviews were conducted on many New York area stations, as well as the Canadian Broadcasting Corporation.

When celebrity sex therapist Dr. Ruth featured teenage environmentalists on her television show, *Ask Dr. Ruth*, Tanja and the other members of the Environmental Club and KAP were the perfect choices for guests.

The special was called "Kids that Get Involved with the Environment," and Tanja, Kurt, and Chriss had to reschedule their midterm exams to appear on the show. It was taped in Queens, N.Y., in front of 150 students from the area and was seen in 40 million homes.

Before the show, Kurt and Chriss got dressed as Ronald McToxic and Ronald McFuture. Tanja's mother and Dr. Ruth Westheimer chatted in their native German when they discovered that they were both originally from that country, and Mr. Stehle ran around taking pictures.

A chemistry professor from Saint Lawrence University in Canton, N.Y., was also there to appear on the show. The professor, Dr. Paul Connet, was also a member of the advisory panel to the Office of Technology Assessment in Washington, D.C.

The show began with Tanja's description of the successful anti-polystyrene campaign they conducted in their schools and the continuing campaign in the town and throughout the country. Tanja also announced their most recent achievement regarding school lunch trays. "West Milford schools will be using washable trays next year because the school board voted to reactivate the dishwashers," she said. "The reusable trays will cost 16 percent less than polystyrene trays. The school district will save $17,328 over paper trays and $6,249 over polystyrene trays." Tanja then held up a plastic tray to the audience and the cameras. "The school also decided to switch to silverware from disposable plastic utensils," she added.

"Will the hard plastic trays be called 'Tanja trays'?" the show's host asked.

Tanja and the others laughed.

Kurt, dressed as Ronald McToxic, then reported the harm that resulted from the use of polystyrene, and Chriss, dressed as Ronald McFuture, described what McDonald's could do to be more environmentally considerate.

Dr. Ruth then turned her questions to Dr. Connet. "If we don't do something right now, what will happen?"

"Due to ozone damage," he said, "agriculture will be harmed by the greenhouse effect. The danger of cancer will increase. Toxics from the manufacture and disposal of plastic products will cause problems with the food chains."

"Are biodegradable plastics the answer?" asked Dr. Ruth.

"No," said the chemist. "Manufacturers should make reusable products. The world can no longer be run as a 'throw away' society. Young people can change the 'throw away' attitude."

Tanja agreed and, taking a cue from Ronald McFuture, concluded the show on a positive note, "If each student would do a fraction of what we do, these problems could be solved."

After being interviewed so many times on television, Tanja had gotten used to the star treatment—make-up, rehearsals, and recognition. She had even been transported by chauffeured limousine more than once from her home to television studios in New York, almost two hours away.

But it was when the cameras started following her around the school that she began feeling uncomfortable with the other students, even her own friends. A number of the TV shows sent camera crews to West Milford High School to tape a typical "Tanja day" and record a sample Environmental Club meeting.

It was true that at television and radio studios and during interviews for magazine articles she was treated specially, but at these times she wasn't in a setting where she was surrounded by her peers. Even though her co-activists from the Environmental Club and KAP often attended and participated in these interviews, it wasn't the same as being singled out from the high school's entire student body.

No, being followed around by a camera in her own high

school was different. She was the center of attention. She was elevated to celebrity status while the other students were reduced to simply background. Not only did she sense that some of the other kids became jealous of her, the comments she overheard made her sure of it.

One time, she even heard a couple of kids she didn't know say, "Now she's going to host her own environmental show." The rumor was simply not true.

Even Mrs. Vogt knew that her environmentalist daughter was becoming alienated by the same girls who had once been her best friends. She often called middle school teacher Norma Sullivan to discuss the change in her daughter's life-style and social life, and Tanja, unaware of these calls or her mother's concern, simply continued doing what she felt was the right thing. She continued writing letters, campaigning against polystyrene, and being sought after by the media, which left very little time for a social life anyway.

Even the Mickey Mouse Club taped one of Tanja's school days. It was January, 1990, a cold winter day, and Disney's camera crew arrived early to set up for a full day of filming. Their breath turned white in the frigid air as they carried their loads of equipment from their trucks to the school. They had arrived early so they could film the Environmental Club having breakfast on the now world-famous cafeteria trays.

By this time, though, the trays were not disposable paper anymore. They were the reusable type that Tanja had first introduced on Dr. Ruth's show. These trays were even more environmentally correct than disposable paper trays, because they were never disposed of in the burgeoning landfills. They were simply washed and reused again and again. Paper use was reduced, so trees were also saved. And the school was happy because using these trays actually cost less. The switch to reusable trays had been inspired by the school's business administrator, Ed Vogel, the same man who had said, "Put your money where your mouth is." After investigating the cost of starting up the school's industrial dishwashers that had been shut down during the energy crisis and oil embargoes of the 1970s, he discovered that it was actually cheaper to use washable trays instead of disposable trays, whether they were paper or polystyrene. So, when Disney came to West Milford, they filmed Tanja and her classmates eating from reusable plastic trays.

After filming breakfast, the Disney camera crew followed Tanja and the other club members throughout the day. As a joke, one cameraman even started following Tanja into the girls' room. "No, no, you can't follow me in there," teased Tanja, going along with the joke. After school, the Disney team filmed the students conducting an Environmental Club meeting and collecting aluminum cans from local stores. The kids demonstrated how they prepared the cans and described how the money raised from recycling was collected to fund a club project to ban animal testing.

The following month, Tanja was even invited to appear in person on the Mickey Mouse Club, where she would receive a Mickey Award and be inducted into the Teen Hall of Fame in Disney World. But the day that she and her family were to go to Florida, Tanja caught a cold. She had already postponed the trip a week to coincide with her father's vacation time, and she already had her airline tickets. Since filming was already scheduled for her appearance on the Mickey Mouse Club, she couldn't wait until she recovered from her cold. Sick or not, Tanja, her mother, father, and sister, Christel, set off for Florida and Disney World.

Even though rehearsals didn't begin until the next day, Tanja began practicing her few short lines the night before in the hotel room. One of her lines was the well-known "See you real soon" part at the end of the show when the Mouseketeers sing the spelling of Mickey Mouse's name. They would begin by singing "M-I-C" in unison, and after the letter "C" Tanja was to say, "See you real soon." Then the mouseketeers would finish spelling Mickey's name by singing "K-E-Y," and someone else would say, "Why? Because we like you." The show was over when the mouseketeers sang, "M-O-U-S-E." It's been a tradition on the Mickey Mouse Club since the 1950s.

Tanja was so concerned about saying this one line that she repeated it aloud hundreds of different ways in the hotel room the night before the rehearsal, sneezing between nearly every try. She tried it with an accent on each word in the sentence. She tried it in a sing-songy voice, and she tried it in a monotone voice. She tried it fast, and she tried it slow. Tanja's parents and sister heard it so many times that night they got sick of it, but Tanja was still not satisfied. She lay awake much of the night blowing her nose and repeating "See you real soon" over and over again in her head.

At the rehearsal the next day she met the stars of the Disney Channel's Mickey Mouse Club and did a pretty good job of hiding her runny nose. Even though the Mouseketeers were all younger than Tanja, they were experienced television celebrities with fans all their own. Although she had no interest in the younger boys on the show, there were hundreds of teeny boppers who would have given anything to trade places with her to be next to the teenage idols.

She realized the extent of their fame when she went dancing with them after rehearsal at Disney World's Pleasure Island. All of the kids at the teen dance club knew the Mouseketeers by sight, and some even told Tanja how jealous they were of her. "My God, you're on the show with them. You're so lucky," one even said.

Tanja's cold lingered into the next day, when her appearance on the Mickey Mouse Club would actually be taped, but fortunately nobody watching the show could tell she was sick. During her segment on the show, she told the story she had told hundreds of times before. Then she suggested to the audience things they could do to help. The Mouseketeers presented her with a Mickey Mouse Award, a large golden trophy in the shape of the world famous mouse, and announced her induction into Disney World's Teen Hall of Fame.

Then Tanja presented her message: "For kids, I would like to give them the message that they can stand up for what they believe in, and they shouldn't think that adults won't listen to them just because they're younger. They should just say what they want to say and make a difference."

One Mouseketeer asked, "What can we do to help save the environment?"

"You can start with the three Rs: reduce, recycle, and reuse," answered Tanja. "You can begin by recycling your daily newspaper, and by the end of the year you can say you saved four trees. I recommend that everyone contribute in their own way from taking on their school board to changing their shopping habits. As long as they're contributing, then they're helping the environment."

Although she was not nervous at all repeating her anti-polystyrene saga because she had practiced it so many times before, she nearly froze when the show came to an end and it was time for her one, short, rehearsed line. As the camera moved in for a close-up, she forgot exactly how she had finally decided she would say the four short words and quickly said "See you real soon" with little inflection at all. "That must have been okay," Tanja assumed. The camera facing her was turned off as quickly as it had been turned on and another camera was turned on to show the television audience a different view of the Mouseketeers as they finished their classic Mickey Mouse Club song. As the closing credits rolled, she and everybody else on stage danced to the Mickey Mouse Rap, a more recent addition to the show. Everyone told her she did great.

With rehearsals over and the show taped, Tanja was free to relax with her family and enjoy Disney World and the other theme parks in the Orlando area. But even at the nearby MGM Studios amusement park, Tanja could not escape her anti-polystyrene campaign. Walking through the park, Tanja and her mother came across a line of people waiting to tour a radio station. The disc-jockey was talking to some of the people through a window, and Tanja's mother stepped right up and started bragging about her daughter. "My daughter's here for the Mickey Mouse Club show, ... and blah, blah, blah. She's leading an international campaign to protect the planet from the harmful effects of polystyrene, ... etc., etc., etc.," she said, telling the same story that Tanja had told so many times before. The disc jockey was so impressed that he immediately invited Tanja in for an interview, and the local radio waves spread her story even further.

But it was after the free trip to Disney World and her appearance on the Mickey Mouse Club that the other members of the Environmental Club became jealous enough that they started complaining to Mr. Stehle about the fact that Tanja got all of the publicity, that she got to be on all of the TV shows and in all of the magazines, that she got the free trip to Disney World.

The teacher tried to explain to them that not only did Tanja do as much work as they did for the club's various activities, but she also wrote hundreds if not thousands of letters to people interested in the anti-polystyrene campaign. He explained how she worked nights and weekends just to keep up. And he was right. No matter how many letters she answered there always seemed to be that many more waiting for responses.

Chapter 10

Earth Day

The year 1990 was a turning point for environmentalism. It was the 20th anniversary of the first Earth Day, and with a growing concern for the planet's health, people throughout the world were taking notice. Local, national, and even international events were planned not only for the purpose of raising awareness of the problems affecting the earth but also to demonstrate ways to solve these problems. The newspapers, radio, and television reports were filled with news of events related to the burgeoning environmentalism that would climax on Earth Day, April 24, 1990.

Tanja, the Environmental Club and Kids Against Pollution had their own plans for Earth Day 1990. The township of West Milford itself, being considered a recreational community with hundreds of lakes, watershed property, and acres and acres of natural woodland, was already a model of environmentalism in some respects. It recycled a large portion of its waste long before it became mandatory throughout the state of New Jersey, and many of its residents were themselves activists devoted to protecting the water and natural land from pollution and development. The township also sponsored its annual Beautification Day, a day when residents would collect garbage from along the roadsides throughout West Milford. This year, Beautification Day was scheduled to coincide with Earth Day, and Tanja and her friends planned to participate. Along with many concerned residents, Tanja and the other enviro-kids planned to walk along the rural roadsides, filling garbage bags with all sorts of debris found along the road.

But just one week before the day, Tanja's plans changed. She got a call from Washington, D.C., inviting her to participate in an Earth Day event that the entire world would be watching. A day-long rally was scheduled to be held on the steps of the U.S. Capitol. Hundreds of thousands of people were expected to attend, famous movie stars and musical performers would appear to perform and to discuss the plight of the environment, and Tanja would be among them.

Her mother reacted as usual when Tanja told her the good news; she got very excited and called all her friends. Then, when her mother finally got off the phone hours later, Tanja called her father at work to tell him, and he also reacted as he usually did. "Oh. Well, that's great," he said. But Tanja was tired of his calm reactions to her exciting news.

"Get excited, Dad," she said. "Don't you understand? I'm going to be involved with Earth Day 1990. It's all over the news, and I'm going to be in Washington, D.C., with a whole bunch of stars, and there will be thousands of people there, and CNN will cover the whole thing live, so millions more people will watch it on TV, all over the world. Dad, this is probably the biggest thing that's happened to me since I spoke at the U.N."

But all her father could muster was "That's great, Tanja. Really, that's great" There was no changing him.

But he was excited enough to drive her the five hours to Washington, D.C., where the Earth Day Committee gave them a room in the Sheraton Carlton Hotel the night before the big day. All of the stars were staying there, and Tanja was hoping to see some of them.

It wasn't long after they checked in that they encountered their first star in the elevator on the way to their room. Tanja dared not say anything until she and her parents had gotten off at their own floor and he was safely out of sight. But as soon as she was sure he couldn't hear her, she burst out, "That was Kevin Bacon. That was Kevin Bacon. Oh, my God, we were so close to him. Don't you know who he is?" Of course, Tanja's parents had no idea who he was, but they were amused by their daughter's reaction to him. Although Tanja knew, her parents had no idea that the young man with whom they had just shared an elevator was the star of the movies *Footloose, She's Having a Baby,* and *A Few Good Men.*

"Oh, he's so cute," continued Tanja. "He's so cute even in real life. He looked like he was going to the gym. He was wearing sweats and everything like he was going to the gym. Oh, Dad, will you go to the gym with me? Oh, Dad, please, please, please can we go to the gym?"

"Come on, Tanja, we just got here," said her practical father, dashing her hopes. "Let's get settled in our room first, then we can do a little sight-seeing or relax before dinner. You'll have plenty of opportunities to see him again, and probably even some other movie stars."

"Other movie stars?" wondered Tanja, and an image of Richard Gere popped into her head. She knew he would be there, too.

"Your father's right," agreed Mrs. Vogt. "Let's just get settled."

But Tanja was excited. She just couldn't get settled. "What do you mean I'll have plenty of opportunities to see him again. How many times will I have the opportunity to see Kevin Bacon working out in the gym?"

"You'll probably see him around the hotel again, and you're sure to see him all day tomorrow," answered her mother.

But Tanja couldn't wait, and she just wouldn't give up. Using the same determination she had mustered to fight McDonald's use of polystyrene, she convinced her father to quickly change into his sweats while she did the same. Together they rushed down to the hotel's fitness center only to discover that Kevin Bacon wasn't there. He had either been there and left before they even got there, or he had never even been there at all. Now it was Tanja's father's turn to do the convincing. "I didn't get changed and come all the way down here for nothing," he said. "I'm staying for a workout, and so are you."

Tanja's plan had backfired. Reluctantly, she agreed and worked out with her father instead of catching a second glimpse of Kevin Bacon. It was an hour before they returned to their room to rest before dinner. While she was lounging around their luxury hotel suite Tanja discovered she would, in fact, see the elusive movie star the following day. The Earth Day organizers had left information about the event in each speaker's hotel room, and while reading through it Tanja discovered that Kevin Bacon was actually scheduled to introduce her. "If he's on stage just before me," she thought, "we'll have to meet each other. Oh my God! Maybe he'll even give me a kiss as I walk out onto the stage like they do at the Oscars." Her disappointment quickly changed to anticipation, and she cheerfully prepared to share dinner with her parents in the hotel's restaurant.

As the hostess led them to their table, Tanja was shocked. She spun around and whispered to her mother, "That's him. He's sitting right there." She dared not point, but she motioned with her eyes, and her mother recognized the same young man with whom they had shared the elevator. "If he read the program, he's sure to recognize my name if I tell him who I am," said Tanja.

"I think you're out of luck, Tanja," teased her mother. "It looks like he has a girlfriend." She referred to the lady sharing a table with him.

"I don't care about that, Mom. I just want to meet him."

Meanwhile, as the two talked and gawked, the hostess had reached their table and was motioning for them to come, and Tanja's father was getting impatient behind the two starstruck women. "You guys go ahead," said Tanja as she gently nudged her parents toward their table. "I'll just introduce myself to him."

If it weren't for the fact that they would both be speakers at Capitol Hill the following day with the added common link that they would both share the same stage at the same time, Tanja would not have had the nerve to approach the star. But, since he was introducing her the following day, she was sure he would recognize her name and be happy to meet her. "After all, I'm not just a brainless fan," Tanja thought to herself as she walked toward his table. "In fact, he's the one who'll be announcing me, not the other way around. He'll have to know something about me to be able to introduce me."

But even with her own encouragement, she was still nervous, and her heart pounded in her chest as she approached him. "Hi. Are you Kevin Bacon?" she asked when she had gotten close enough for him to hear.

"Yes," he said, hardly looking up.

"I'll be speaking at the Earth Day Rally tomorrow, too."

"Oh," he said, not offering any information nor requesting any.

"Guess he doesn't care," thought Tanja, feeling her face turn red from embarrassment. "Well, one more try. It's probably my last chance anyway."

"My name is Tanja Vogt," she said hoping he would recognize her name.

But obviously he didn't, because again all he said was, "Oh."

"Well, nice meeting you," she said, as if she had really met him. "I guess I'll see you tomorrow," she said as she slunk toward the table where her parents were sitting.

"I guess he didn't read tomorrow's program," Tanja said to her parents. "He had no idea who I was and didn't even want to know."

But by the next day, Tanja found herself an average person among stars, the only Earth Day speaker who was not a superstar, and everyone wanted to know her story. It seemed that everyone in the hotel was headed over to the rally the following morning. The organizers were telling people where to go and how to get there, and someone ushered Tanja to one of the fleet of vans reserved for bringing the speakers to the U.S. Capitol. There was one seat left, and Tanja could have it. Her parents would have to drive themselves in.

As she approached the van, she couldn't see in because the windows were tinted. But as she ducked her head and stepped into the van, she nearly fell back out when she saw who she would be sharing the ride with. Among the stars seated in the van were actor Richard Gere, super model Cindy Crawford, and singer Olivia Newton John.

Silently, she took her seat. Starstruck, she spent the entire ride in shock, listening to Olivia Newton John and Richard Gere make small talk as if they were typical strangers meeting on a bus or in a supermarket checkout line. Nobody talked to Tanja, and Tanja talked to no one. She was hardly in the habit of striking up a conversation with typical strangers let alone mega-stars with wealth and fame and thousands of admirers eager to just see them in person. She spent the ride in a daze, making it seem instantaneous that they arrived at their destination, the U.S. Capitol and Earth Day 1990, where Tanja would address a crowd of thousands and be seen by a global audience of millions.

Movie star Richard Gere stepped from the van first, and the crowds surged toward the temporary barricades set up to keep them at a safe distance from the stars they had come to see. The barricades served their purpose, and the stars seemed to hardly notice the throngs of fans as Richard Gere helped the others as they stepped from the van, even Tanja. She could hardly believe it when he clasped her hand to ease her descent from the vehicle. "I'm really touching him," she thought, "Richard Gere." Tanja felt like royalty, like a star among stars as their fans clamored to see and take photographs.

Tanja walked along with the others through temporary fences set up to channel them safely passed the hordes of onlookers who had come not only to worship the stars but the earth as well. By now, Olivia Newton John and Richard Gere were beyond small talk and were discussing the day's rally.

"And what will you be speaking about?" Tanja thought she heard someone ask through the fog of her daze. But she ignored it thinking that it couldn't be meant for her. Then she heard it again, a little louder and a little more consciously.

She focused and saw that Richard Gere was speaking directly to her as they walked toward the tents set up to house the speakers for the day. Both he and Olivia Newton John were looking at Tanja waiting for her reply.

"Oh. I'm sorry," she smiled awkwardly. "I didn't realize you were talking to me. I must have been daydreaming."

They both smiled back at her, making her feel more at ease, and she told them the story of her anti-polystyrene campaign, the same story she had told so many times before during interviews for radio, television, newspapers, and magazines. And the entire time they walked from the van to the backstage area, people yelled and waved and took pictures, and Tanja thought to herself, "They must be thinking, 'Who the heck is she?'"

Eventually, Tanja caught up with her parents backstage. After coordinating the time of Tanja's appearance with the organizers of the rally, they were free to wander around, eat the food prepared for the speakers, and meet the many famous people gathered there that day. Tanja got her father to take her picture with nearly everyone she met, including Woody Harrelson, John Ritter, Malcolm Jamal Warner, John Denver, and even Kevin Bacon, who was a lot friendlier when he realized that he would be introducing Tanja later that day. She met the lead singer of REM and the rock band 10,000 Maniacs, which were scheduled to perform immediately after Tanja's speech.

But even when Tanja's father suggested it himself, she was reluctant to get her picture taken with Senator Al Gore, another one of the scheduled speakers.

"Why do I want a picture with him? Nobody knows who he is," said Tanja after her father's suggestion.

"You'll be glad you have this picture when he becomes president or something," said Mr. Vogt.

"Yeah, right," she said, but got her picture taken anyway with the man who was a little known senator back in 1990.

For Senator Al Gore, the "or something" Mr. Vogt predicted eventually came true two years later when he was elected vice president of the United States. As the first real modern environmentalist to make it that close to the White House, his popularity was based in large part on his best-selling book, *Earth in the Balance.*

Tanja got pictures with other stars that day, too, but even though superstar Tom Cruise was also there to speak, he was the most difficult to get close to. He and his entourage took up one entire tent, and his bodyguards stationed at the entrance ensured that nobody uninvited could enter. He didn't even mingle with the other stars.

But Tanja had quickly gotten used to meeting and speaking with famous people that day, and she eventually felt comfortable enough to approach Tom Cruise's bodyguards. "Hi, I'm Tanja. Can I say hello to Tom Cruise?" she asked one of his bodyguards.

"No," was the answer, firm and final. The fact that Tanja had become comfortable introducing herself to the other speakers did nothing to change the answer. She walked away disappointed.

Soon after, she saw a woman stepping out of the tent that Tom Cruise had claimed for his own. Tanja thought she recognized the woman but couldn't immediately place her face. Then it hit her. "That's the lady from *48 Hours*," she told her father. The woman, Bonnie Reese, organized a group of famous environmentalists, and Tanja had seen her on the television news magazine show known as *48 Hours* only the week before.

"Yes, that's me," said the woman who had overheard Tanja. "And who are you?"

"I'm Tanja Vogt. I'm speaking today."

"Tanja Vogt. Yes, I heard your name. I heard you would be speaking today. What will you be speaking about?"

Again, Tanja told the story that she had told so many times before. After Tanja and Bonnie talked a little more, Bonnie asked, "Have you met Tom?"

"Tom?" Tanja wondered for only a split second before she realized that Bonnie meant Tom Cruise.

Tanja smirked as she walked with Bonnie past the bodyguard who had refused her entrance only minutes before.

She met Tom Cruise and got her picture taken with him. He told her that as a child he had actually lived in Bloomingdale, one of the towns adjacent to West Milford, and Tanja told about her anti-polystyrene campaign for the umpteenth time that day.

By midday, the crowd in front of the U.S. Capitol had grown to 500,000, and millions more watched the rally live on CNN. Eventually, Kevin Bacon introduced teenage environmental activist Tanja Vogt, and Tanja walked out to the microphone. Kevin Bacon didn't kiss her, but that didn't matter to her anymore.

The enormous crowd, like the audience of a sold-out outdoor concert, were packed in between the stage on which she stood and the Washington Monument. The white marble monument pointed toward the sky, the sun, and the deteriorating ozone layer, Tanja's evidence of an earth being weakened by the actions of humanity. In contrast to the monument pointing toward the heavens, the thousands of people gathered together to worship the earth and the environment of which they were an integral part represented humankind's ability to correct the harm that it had caused. Tanja was there to reassure them that it could be done.

She looked out over the mass of people, fused as a whole from her vantage point as if they were a single organism, and she told them her story. She told them that one person could make a difference, had already made a difference, and would continue to do so. The crowd cheered in response to many of her achievements. She never referred to the notes she had brought with her, tucked in her palm, in case she went blank in front of the microphone. She knew exactly what she wanted to say:

"I stand before you today because my efforts have been noticed and have gotten a lot of attention, but there are many people who are doing just as much to clean up the environment and never get noticed. I hope that today will be the start of a new trend so that some day we can look back at the '90s as a decade of change.

"The key to cleaning the environment is the three Rs: reduce, reuse, and recycle. I must admit I get depressed, too, when I think about all that pollution and about all those people who just don't care. But now I see all of you here today ready to make a difference, and I have to say the future looks brighter than ever."

She thanked them for being there, for caring, for making a difference, and she left the stage. The crowd cheered. And as soon as Tanja finished speaking, her father uncharacteristically screamed out as loud as he could, "Alright, Tanja," and his enthusiastic message was picked up by the CNN microphones and broadcast for all the world to hear.

◦◊ ◦◊ ◦◊

Soon after she returned from telling the world about her anti-polystyrene campaign at Earth Day 1990, Tanja went back into action at another McDonald's rally. It was April 29, 1990, one full year after the Environmental Club's first "Send-It-Back" rally. By now it was clear that the corporation was not going to keep to the pledge that Tanja had presented only five months earlier at a similar rally held at the same McDonald's. The Earth Day 1990 deadline had come and gone, and McDonald's had not achieved all four requirements of the pledge handed directly to the corporation's representative, Judy Donaldson, in November of 1989.

Now it was time for the Environmental Club, Kids Against Pollution, and hundreds of thousands of other environmentally concerned citizens to live up to their side of the pledge—to boycott McDonald's. Only five months earlier, Tanja had said, "They have five months to get their recycling efforts going. If the restaurants do not begin recycling 50 percent of their polystyrene by Earth Day in April, we're going to call for a boycott of all McDonald's products." It was time to begin that boycott.

Unfortunately, even though this was the most important message that Tanja and her fellow environmental activists had presented to the corporation up until this point, not a single representative from McDonald's attended the rally—not the corporate executives who had willingly flown out to West Milford before, not the district manager who had personally listened to the speeches made at previous rallies, and not even the franchise owners whose property was the site of the current protest. Instead, an unnamed official from the restaurant emerged to ask the protesters to move to a corner of the parking lot.

So they did, and there, in the presence of a growing group of anti-polystyrene activists, some that had come from as far away as Vermont, Tanja announced that the McDonald's boycott would begin.

As the cameras rolled and the journalists wrote down her words, Tanja said, "No one from McDonald's would listen to us, and since no one will listen to our verbal pleas, we will now start affecting them economically by boycotting their restaurants. If they continue to use polystyrene, they are going to lose a lot of money." Mr. Stehle was there to help reinforce Tanja's message. Wearing an anti-pollution T-shirt and a camera slung around his neck to record the milestone event, he walked around carrying a plastic toy cash register to illustrate that if McDonald's continued using nondegradable polystyrene they would soon feel the effects economically.

The crowd cheered. Then Mr. Stehle made his own announcement. "Kathy Hinds, who heads the National Toxics Campaign with over 100,000 members, has also announced a nationwide boycott against McDonald's."

The crowd cheered again, and then Mr. Stehle added to the growing reasons for boycotting the use of polystyrene. "The bad news about polystyrene just keeps coming in," he said. "Last Tuesday, scientists recognized that hydrohalocarbons—that's a chemical component of polystyrene—are 410 times more powerful as heat trappers than carbon dioxide. With this information, some scientists at DuPont determined that hydrohalocarbons are themselves directly responsible for 5-10 percent of the greenhouse effect. So, not only is polystyrene breaking down the ozone layer and causing low level pollution, but it also adds to the greenhouse effect. We've got to get rid of this stuff."

The crowd cheered again, and then started chanting "Styrofoam has got to go. Styrofoam has got to go."

And even though nobody from McDonald's would hear the message in person, there were plenty of television cameras and newspaper reporters to carry the message to the public and back to the executive offices of the McDonald's corporation. They would hear soon enough.

Chapter 11

Sticking her neck out and into Russia

Tanja's appearance on Earth Day, her announcement of a nationwide boycott of McDonald's, and the monthly anti-polystyrene rallies that continued even after the boycott announcement not only generated widespread interest in her campaign but also enhanced her personal recognition as well. Magazine interviews and television appearances continued, and the letters kept pouring in.

One day a letter arrived from The Giraffe Project, an organization that honors people who "stick their necks out" for a cause. The project was sponsored by Ben & Jerry's, an environmentally conscious ice cream company. Along with 30 other 9- to 19-year-old "giraffes" from 12 different states, Tanja was invited to join in a tour of cities throughout Russia. While most of the other kids on the trip were also environmental "giraffes," some, such as the boy who worked with people afflicted with Down's Syndrome, stuck their necks out for other causes.

The trip started in Burlington, Vermont, on June 23, 1990. In Vermont, the group toured the factory operated by Ben & Jerry's. The company is well known for its Rainforest Crunch ice cream made with nuts grown in endangered rainforests, and Cherry Garcia ice cream, named for Jerry Garcia, lead singer of The Grateful Dead and also an environmental activist working to help save the rainforests.

From Vermont, the group flew to New York, where they got on another, much longer flight to Helsinki, Finland, gateway to Russia. And finally, after a 16-hour train ride, they arrived in Moscow. It was during the long train ride, packed four to each small compartment, that the "giraffes" got to know each other, discussed their causes and shared their feelings, at first about their expectations of Russia and eventually about themselves.

"Well, we're really on our way now," said one of the other girls in Tanja's compartment. "I can't believe I'm actually on a train to Russia."

The others voiced and nodded agreement. "I wonder what it will be like," said Tanja, carrying on the conversation.

"It'll probably be like that commercial. You know, the one with the fat women modeling Russia's latest swimming fashions, and they're all the same ugly grey bathing suits," joked one boy.

They all laughed because they all recognized the commercial he was talking about.

"You know," continued the same boy, "anything you ever see about Russia shows old people bundled up against the cold in dingy black or brown overcoats and waiting in line for scraps of food. And all the buildings around look like something out of a fairy tale or the top of a multicolored Dairy Queen ice cream cone."

They all laughed again because they all had the same image. Their shared expectations of Russia were all based on the brief glimpses they had seen at home on television or in magazines.

"Did everyone bring their presents?" Tanja asked, changing the subject to a more positive aspect of the foreign culture they were visiting. The sponsors of The Giraffe Project had told the kids that Russian people often give their own prized possessions to someone they like to ensure that the friendship will last. The sponsors had suggested that the kids bring their own gifts so they could reciprocate if necessary.

"Yup, got 'em in my suitcase," said the boy who had been joking about the grey Russian life-style, and the others said that they had also brought presents. "And for a little trading," continued the same boy, "I brought some jeans and sweatshirts. The Russians love 'em, but they're too expensive over here. I'll give them some typical American clothes in exchange for some typical Russian items."

They spent the entire 16 hours talking and eating and sleeping and squirming to find a comfortable position on the Russian train. When the young activists finally arrived in Moscow, it wasn't nearly as dark, as cold, or as grey as they had expected.

But the kids were tired and worn out from the 16-hour train

ride when they unfolded themselves from their cramped positions and trudged from the train. As the first among them stepped onto the platform, they were surprised by the warm greeting they received from a group of children and adults waiting for them at the Moscow train station. The welcoming committee was prepared, and welcomed the group with songs, flowers, presents, and even a Russian dance.

The rousing welcome briefly perked up the weary group, keeping them cheerfully awake long enough to check into their hotel rooms and collapse on their beds. And for their first night in Russia, they saw nothing but the sheets. It wasn't until the next day that they got their first good look at Moscow, starting with their hotel rooms, which they soon discovered were nothing like the hotels they had stayed in back in the United States.

Some rooms had tubs with old-fashioned basins standing atop iron feet. Other rooms had no tubs at all but simply showers with no shower curtains. And all of the rooms lacked another modern convenience, soft toilet paper; it was more like hard paper towels. The room Tanja shared with another girl was one with a shower, but its lack of a shower curtain confused her. "How am I gonna do this?" she wondered.

She eventually figured it out and showered, but drying her long hair was impossible. She had brought her blowdryer and an adapter that was supposed to work in the foreign electrical outlets, but it didn't. So Tanja spent the first half of her first day in Russia walking around with a wet head. The rest of the group were suffering from their own culture shock, and they all immediately agreed that the food was bland, at best. Even in the best hotels, orange juice and milk were not available.

But compared to the natives they saw everyday, the kids weren't so bad off. For their own daily requirements of food, clothing, and other items, the Russians themselves spent a large portion of the day standing in line.

The group traveled to three Russian cities—Moscow, Kiev, and Leningrad (now known as St. Petersburg). They spent three or four days in each city, where they met with their Russian counterparts, kids who were also sticking their necks out for their own communities.

By getting together and sharing stories about the causes that really mattered to them, the kids learned not only about the country they were visiting but also about themselves. They learned songs about the environment, and both the Russians and the Americans sang them together as a group. The Russians loved hearing the Americans sing so much that they asked them to sing over and over again.

And sharing time with each other, they came to realize that the people they had originally thought of as dreary and oppressed actually turned out to be among the world's most generous, starting from even the youngest members of a society that was no longer unfamiliar to this group of American kids.

The boy who had first joked on the train about the "old, fat people bundled in overcoats" spent most of his time at a picnic one day playing with a five-year-old boy. By the end of the day, the boys had become so close that the five-year-old gave the older boy his favorite teddy bear. The generosity made the older boy cry, but he dared not risk insulting his new friend by refusing the gift.

On the trip home, the conversations were different than they had been at the beginning of their journey. Their images of Russia were now based on reality and not on some limited images they had seen on television or in magazines. "You know, we probably have more clothes in our suitcases than those kids have in their entire wardrobe," said one girl. "Did you notice that most of those kids wore the same clothes every day, and we had enough clothes in just our suitcases to have a change of clothes every day."

"Yeah, and they have to wait in line all day just to get the little bit that they do have," another one said.

Both the chaperones and the kids cried when they parted at the end of the trip, and some planned reunions that never took place. But they all had new Russian pen pals to write to in the coming months, and soon after returning home, one boy with a changed image of Russia mailed his own childhood teddy bear to the five-year-old boy who had demonstrated Russian generosity.

Chapter 12

McDonald's makes an announcement

The anti-polystyrene rallies continued soon after Tanja returned from Russia. Then it was time for school to start and another full year of the Environmental Club, which planned a schedule of monthly rallies. One rally was scheduled for October 30, 1990, a full six months since the McDonald's boycott first began.

The rallies had changed some since they had begun about a year and a half before. Chriss had long since lost interest in her part as Ronald McFuture which was taken over by Laura Moskin, Lisa Care, and other members of the Environmental Club. Kurt had been getting into trouble and could only play Ronald McToxic when he wasn't suspended from school. Still, with other New Jersey schools joining in, there were many more participants than during the first few rallies. But Kurt was suspended on October 30th, the day of the scheduled rally, and some of the schools that had originally planned to be there could not participate, so the rally was postponed until the next month. It would never be held.

On November 1, 1990, two days after the last scheduled rally, the high school's vice principal called Mr. Stehle down to his office. And, just as two years before, Tanja had worried that Mr. Stehle had written "See me after class" on her current events assignment, Mr. Stehle himself wondered why the vice principal wanted to see him. "What did I do?" he thought as he walked toward the high school offices. "I turned my lesson plans in. Why does he want to see me?"

Soon after, Mr. Stehle called Tanja out of her math class and asked her to report to the vice principal's office. Now it was Tanja's turn to wonder what she did wrong. She couldn't think of anything.

As she approached the office, Mr. Stehle's impatience was apparent. He was waiting in the hall for her, and when he saw her coming he called the vice principal out into the hall with him. He could see her down the hall, but couldn't wait for her to get any closer. "Tanja!" he called, "They switched! They're going to switch! McDonald's is going to switch!"

Tanja couldn't believe it. At first, she stopped walking; then she ran into Mr. Stehle's arms and gave him a hug. She even hugged the vice principal who she hardly knew.

"That's great! I can't believe it! That's great!" she said, ecstatic. "How did you find out?"

Mr. Stehle pointed his thumb at the vice principal, and the vice principal said it was in the newspaper.

The New York Times carried the story in its November 1, 1990 issue, almost two years exactly since the day Tanja had clipped her current events article from the *West Milford Argus*. The *Times* article reported that McDonald's announced that within 60 days it would begin eliminating from its 8,200 restaurants all plastic-foam sandwich containers, accounting for nearly 75 percent of the company's total foam use.

McDonald's president, Edward Rensi, was quoted in the article as saying, "Although some scientific studies indicate that foam packaging is environmentally sound, our customers just don't feel good about it. So we're changing."

"It finally paid off!" said Tanja, hugging Mr. Stehle again.

The high school's switchboard was lit up with calls from newspaper and television reporters hoping to get quotes from the girl who persuaded McDonald's to care about the environment.

And although she had little time to understand her own feelings, she answered the reporters as well as she could: "It feels great to know that I was in some small way responsible. It shows that the consumer can really make a difference. If you stick to something and get a lot of people together, you can make an impact. It's been such a long effort, but it was worth every minute of it now. I just did my homework one night, and it snowballed into this huge campaign. And now this major company is doing something about the environment."

Mr. Stehle had his own answers for the telephone interviewers: "We are overjoyed today at this announcement. This serves as a valuable lesson to the students; it shows them that they can win even against tremendous odds. McDonald's did a responsible thing today. They are trying to be responsible because no one wants to pollute the environment. As a matter of fact, some of the club members said they might start eating at McDonald's again. Don't worry, though. We'll always be working on something because the environment needs a great deal of help from all of us."

A number of television news programs weren't content with a simple quote taken over the telephone; they wanted footage of Tanja and Mr. Stehle and the other anti-polystyrene enviro-kids reacting to McDonald's announcement. So they immediately sent crews to the West Milford High School to film the response of the successful environmental activists.

And while the media were on their way and in between conducting telephone interviews, Mr. Stehle and Tanja also called Kurt out of class and down to the vice principal's office. He cried when he heard the news, and through his tears of joy, he said, "What do we protest next?"

With Tom Cruise

With Kevin Bacon and Kyra Sedgewick

EARTH DAY 1990

With Shanna Reed

With Richard Gere

Tanja giving her speech at the United Nations

Thanks for reading my story. If you'd like to join in on making a difference in the environment, check out the following organizations and resources.

Tanja Vogt

APPENDIX

Organizations Of Young Environmentalists

Isabel S. Abrams
Caretakers of the Environment International/USA
2216 Schiller Ave.
Wilmette, IL 60091
708-251-8935

**Children's Alliance for Protection of the Environment
(CAPE)**
P.O. Box 307
Austin, TX 78767

Children's World Seminar
Ellen Brogren
150 Bella Vista Dr.
Hillsborough, CA 90401
415-342-2579
Fax:415-349-5862

Conservation Career Development Program
Benjamin Luke, Prog. Assist.
1800 North Kent St, Suite 1260
Arlington, VA 22209
703-524-2441
Fax: 703-524-2451

EarthDance Institute
Mark P. Fields
PO Box 2155
Asheville, NC 28802
704-252-8188
Fax: 704-258-1418

Earth Education Project
Mr. Mark Gerzon
2218 21st St.
Santa Monica, CA 90405
213-450-9002
Fax: 212-362-9847

Earth Train
99 Brookwood Road, Suite 3
Orinda, CA 94563-3344
510-254-9101
Fax: 510-254-8720

Ground Truth Studies Project
The Aspen Global Change Institute
100 East Francis
Aspen, CO 81611
303-925-7376
Fax: 303-925-7097

KiDS STOP
KiDS Save The Planet
P.O. Box 471
Forest Hills, NY 11375

Kids Against Pollution
Tenakill School
275 High Street
P.O. Box 775
Closter, NJ 07624
201-768-1332

Kids For a Clean Environment (FACE)
P.O. Box 158254
Nashville, TN 37215

Kids for Saving the Earth (KSE)
Teresa Hill, President
P.O. Box 47247
Plymouth, MN 55447-0247
612-525-0002

Kids Network
National Geographic Society
Educational Services, Dept. 1001
Washington, DC 20077
800-368-2728

Peace Child International
Mr. David Woollcombe
Maltings Lane
Much Hadham
Herts. SG10 6AW
United Kingdom
011-44-279-842-414
Fax: 011-44-279-843-428

**People Educating Other People for a
Long-Lasting Environment (Project PEOPLE)**
P.O. Box 932
Prospect Heights, IL 60070

**Rainforest Education Leadership and
        Rainforest Scholarship Fund**
(see EarthDance Institute)

Student Conservation Association
1800 N. Kent St., Suite 1260
Arlington, VA 22209
703-524-2441
Fax: 703-524-2451

Regional Offices:
Rob Bleiberg
SCA SW Regional Coordinator
PO Box 925
Boulder, CO 80306
303-938-6834
Fax: 303-938-6850

Rueben Duenas
PO Box 151295
Los Angeles, CA 90015
213-747-6798
Fax: 213-749-3331

Pete Sanborn
SCA NW Regional Coordinator
2524 16th Ave., S.
Seattle, WA 98144
206-324-4649
Fax: 206-324-4998

Robert Robinson
PO Box 32369
Newark, NJ 07102
201-733-4450
Fax: 201-733-4452

Ray Auger
PO Box 550
Cahrlestown, NH 03603-0550
603-543-1700
Fax: 603-826-7755

Duane Poe
655 13th St., Suite 304
Oakland, CA 94612
510-832-1966
Fax: 510-832-4726

Student Environmental Action Coalition (SEAC)
P.O. Box 1168
Chapel Hill, NC 27514-1168

United Nations Environmental Programme
      Global Youth Forum
Ms. Dulcie de Montagnac
United Nations
Room DC1-590
1 UN Plaza
New York, NY 10017
212-963-7781/4931
Fax: 212-963-8193

Youth Action
1830 Connecticut Ave., NW
Washington, DC 20009
202-483-1432

Youth Environmental Service Awards
(see EarthDance Institute)

Youth Environmental Service Computer Network
      (YES NET)
P.O. Box 2155
Asheville, NC 28802
704-252-8188
Fax: 704-258-1418

Youth Environmental Society
P.O. Box 441
Cranbury, NJ 08512
609-655-8030

Youth for Environmental Awareness
Diane Danielson, adult coordinator
3204 Northwest 27th Ave.
Portland, OR 97212
503-284-5377

Youth for Environmental Sanity (YES)
Ocean Robbins
706 Frederick St.
Santa Cruz, CA 95062
408-459-9344
Fax: 408-458-0255

Miscellaneous

Eco-foam
A starch-based substitute for polystyrene packing peanuts—decomposes in compost heap and under running water
>Ken Starrett
>Marketing Department
>American Excelsior Co.
>850 Avenue H East
>P.O. Box 5067
>Arlington, TX 76005-5067

Books For Young Environmentalists

50 Simple Things Kids Can Do to Save the Earth
Animal Town
P.O. Box 485
Healdsburg, CA 95448
$7.00

365 Ways For You and Your Children to Save the Earth One Day at a Time
by Michael Viner with Pat Hilton
Dove, Inc.
Warner Books, Inc.
666 Fifth Avenue
New York, NY 10103
212-484-2900

Adventures of the Garbage Gremlin
free comic book for grades 4-7
Office of Program Management/Support, U.S. EPA
OS-305, 401 M St. SW
Washington, DC 20460
Order #EPA/530-SW-90-024

The Big Green Book
by Fred Pearce
Grosset & Dunlap
$13.95

Creating Our Future Manual
Student Action Group Manual
Creating Our Future
398 North Ferndale
Mill Valley, CA 94941
415-381-6744
$3.00

E for Environment
annotated bibliography of 517 children's
          environmental books
by Patti K. Sinclair
R.R. Bowker
P.O. Box 31
New Providence, NJ 07974
800-521-8110
$44.70

Kid Heroes of the Environment
The Earthworks Group
Earth Works Press
1400 Shattuck Ave. #25
Berkeley, CA 94709
$4.95

The Kids Environment Book
by Anne Pedersen
John Muir Publications
P.O. Box 613
Santa Fe, NM 87504
$13.95

My Earth Book
The Learning Works
for ages 6-9
puzzles, projects, and pictures to color
$7.95

**My First Green Book: A Life-Size Guide to Caring
for Our Environment**
Dorling-Kindersley, $12

Projects for a Healthy Planet:
    Simple Environmental Experiments for Kids
by Shar Levine and Allison Grafton
John Wiley & Sons, Inc.
Profess. Reference and Trade Group
605 Third Ave.
New York, NY 10158-0012

The Student Environmental Action Guide:
    25 Simple Things We Can Do
The EarthWorks Group
EarthWorks Press
1400 Shattuck Ave. #25
Berkeley, CA 94709
415-841-5866

Teaching Kids to Love the Earth
Animal Town
P.O. Box 485
Healdsburg, CA 95448
$15.00

Earth Book For Kids: Activities to Help Heal
the Environment
(grades 4-6)
by Linda Schwartz
The Learning Works $9.95
P.O. Box 6187
Santa Barbara, CA 93160

Going Green
by Catherine Von Ruhland
Marshall Pickering
HarperCollinsPublishers $5.95 ($7.95 Canada)

Magazines for Young Environmentalists

Bear Essentials News for Kids
P.O. Box 26908
Tempe, AZ 85283
602-345-7323

Family Earth
129 West Lincoln Ave.
Gettysburg, PA 17325

How on Earth!
Teens supporting compassionate, ecologically sound living
P.O. Box 3347
West Chester, PA 19381
717-529-8638

Owl Magazine
(ages 8-14)
56 The Esplanade,
Ste. 306
Toronto, Ontario M5E 1A7
Canada
416-868-6001

The Record's GeoClub
(ages 15 or younger)
GeoClub
Sunday Living
The Record
150 River St.
Hackensack, NJ 07601

Awards For Young Environmentalists

Environmental Exchange
1930 18th St. NW
Suite 24
Washington, DC 20009

Giraffe Project
P.O. Box 759
Langley, WA 98260
206-221-7989
Fax:206-221-7817

President's Environmental Youth Awards
401 M St. SW
Washington, DC 20460

Searching for Success
Renew America
attn: N. Grimes
1400 Sixteenth St. NW
Washington, DC 20036
202-232-2252
Fax: 202-232-2617

Windstar Youth Award
2317 Snowmass Creek Road
Snowmass, CO 81654
303-927-4777
Fax: 303-927-4779

More titles for Young Adults from
The Book Publishing Company

Kids Can Cook	$ 9.95
Dream Feather	$11.95
Nature's Chicken	$ 5.95
The Story of Today's Chicken Farms	
A Natural Education	$ 5.95
Sacred Song of the Hermit Thrush	$ 5.95
Song of the Seven Herbs	$10.95
Song of the Wild Violets	$ 5.95
Spirit of the White Bison	$ 5.95

Other Environmental Books from
The Book Publishing Company

Climate in Crisis:	
The Greenhouse Effect and What We Can Do	$11.95
Ecological Cooking	$10.95
How Can One Sell The Air?	$ 6.95
No Immediate Danger	$11.95
Shepherd's Purse: Organic Pest Control Handbook	$ 9.95
Shopping Guide for Caring Consumers:	
A Guide to Products Not Tested on Animals	$ 5.95

Ask your store to carry these books, or you may order directly from:

The Book Publishing Company
P.O. Box 99
Summertown, TN 38483

Or call: 1-800-695-2241
Please add $2.00 for shipping